Sailing with Noah

Praise for *Sailing with Noah*

"Noted conservationist Dr. Jeffrey Bonner shares insight from his career with animals in this new book entitled *Sailing with Noah*. Through his amazing experiences with animal husbandry in zoos to animal behavior in the wild, he takes a unique look at how conservation should be approached, as we humans are—indeed—the keepers of our world!"

—JACK HANNA, Director Emeritus, Columbus Zoo and Host, *Jack Hanna's Animal Adventures*

"Jeff Bonner is a natural storyteller with a marvelous eye for the meaningful. This is important because what he has to say is important, often unexpected, and even fearless. His insights on the human-wildlife relationship are presented so engagingly that his points penetrate deeply, whether about the sensibleness of some gorillas or the senselessness of some animal-loving humans. This is easily the best 'zoo book' I have ever read, partly because it is about so much more than zoos. It is personal, enjoyable, and also kindly. If you are interested in animals, you will enjoy the revelations and take hope from Bonner's optimism."

—WILLIAM CONWAY, Senior Conservationist and former President, Wildlife Conservation Society

"*Sailing with Noah* is a highly readable, enjoyable, and engrossing account of the turnaround zoological parks have made to meet the impending crisis of nature's disappearance. Ever since Gerald Durrell's and James Herriot's deaths a decade ago, there has been an unfilled niche in the market for animal lovers. Jeff Bonner takes that genre a leap into the modern age. . . . Bonner's chatty, earnest, direct style is convincing and powerful. This book will have profound influence."

—PATRICIA C. WRIGHT, Executive Director, Institute for the Conservation of Tropical Environments

"In his easy-going, delightful prose, Jeff Bonner takes us gently on a journey that leads from pandas to philosophy, spitting cobras to rhinos and Amur tigers, burying beetles to ruffed lemurs and sifakas, Madagascar to Botswana, the Madagascar Faunal Group to PETA, but it is really a journey into ourselves, the symbolism and reality of zoos in our lives, and a worthy challenge to us to take care of our troubled planet. Zoos as a safety net for species endangered in the wild, zoos as a place where people can meet the emissaries of other species and come to appreciate them, zoos as a metaphor for the whole world—Bonner thoughtfully discusses each of these, weaving a rich account of zoos in the modern world, of his own life, and of our individual obligations. *Sailing with Noah* is a book to be read by anyone who values the present or cares about the future."

— PETER H. RAVEN, Director, Missouri
Botanical Garden, St. Louis

"The time is long overdue for a straightforward account of zoos, what they do, how they do it, and probably most important of all, why they do it. Jeff Bonner's *Sailing with Noah* effectively fills this niche and more besides."

—DAVID MORGAN, Executive Director, African
Association of Zoos and Aquaria

Sailing with Noah

Stories from the World of Zoos

Jeffrey P. Bonner

UNIVERSITY OF MISSOURI PRESS   COLUMBIA AND LONDON

Library of Congress Cataloging-in-Publication Data

Bonner, Jeffrey P.
  Sailing with Noah : stories from the world of zoos / Jeffrey P. Bonner.
     p. cm.
  Summary: "An intensely personal, behind-the-scenes look at modern zoos,
written in a lively, accessible style. Through a variety of true stories, some
funny, some sad, occurring in different cities and on different continents,
Bonner describes the changing role of zoos and argues that conservation is
the shared responsibility of all mankind"—Provided by publisher.
  Includes bibliographical references (p.   ) and index.
  ISBN-13: 978-0-8262-1636-6 (hardcover : alk. paper)
  ISBN-10: 0-8262-1636-6 (hardcover : alk. paper)
  ISBN-13: 978-0-8262-1637-3 (pbk. : alk. paper)
  ISBN-10: 0-8262-1637-4 (pbk. : alk. paper)
  1. Zoos—Anecdotes.   I. Title.
  QL76.B66 2006
  590.73—dc22

                                        2005031401

Designer: Kristie Lee
Line illustrations: Maili Poag
Typesetter: Crane Composition, Inc.
Printer and binder: Thomson-Shore
Typefaces: Berkeley and Novarese

This book is dedicated to my wife, my mother, and my son. All have inspired me, each in different ways.

Contents

Acknowledgments

Common sense would tell you that waiting until the day a manuscript is due at the publisher's and then trying to do the acknowledgments is not a good idea. Sometimes, however, schedules and common sense collide. Not only am I waiting until the last possible minute to thank all of the people who helped me prepare this manuscript, but I'm also trying to pack for a trip to Madagascar. I'd prefer to overlook one or two essentials for the trip in favor of not overlooking one or two people who really helped and whom I might forget. But I'm pretty sure I'll manage to do both.

Many people helped write this book. Some of them helped by reading the manuscript, rolling their eyes, and shrugging. Others helped by finding many of the seemingly infinite number of mistakes I made and suggesting that perhaps I should, to paraphrase Mark Twain, get my facts straight first, before I distort them. To all of you—the eye rollers and the fact correctors alike—I must stress that the remaining mistakes are my own. You should have seen it before your corrections were made. Believe me, it's much better now.

I'd like to thank Jim Alexander, Cheri Asa, Andy Baker, Jonathan Ballou, John and Cynde Barnes, Steve Bircher, Lew Birmelin, Rich Block, Bill Boever, Onnie Byers, Bill Conway, Ellen Dierenfeld, Martha Fischer, Judy Gagen, Ron Goellner, Paul Grayson, David Hagan, Charlie Hoessle, Pete Hoskins, Terri Hunnicutt, Alvin Katzman, Devra Kleiman, Lee Lyman, Tracey McNamara, Jerry McNeal, Tom Meehan, Dave Merritt, Bob Merz, Eric Miller, Ginny

Morgan, Debbie Olson, Paul Pierce-Kelly, Ingrid Porton, Jeff Proudfoot, Melinda Pruett-Jones, Myrta Pulliam, Jan Ramer, Lee Simons, Jane Stevens, and Meg White for giving me their stories or helping me to keep my stories straight. Patrick Boddam-Whetham, Dave Morgan, and Pat Wright reviewed the entire manuscript and provided advice and encouragement, as did my wife's book club. My brother Roger read an early version and thought it was pretty good. Beyond that, he didn't help much.

Lana Jordan (the granddaughter of George Vierheller) provided me with her grandfather's unpublished manuscript and allowed me to reproduce large sections of it. Jill Gordon, Christie Childs, and Kevin Kampwerth helped me with citations and photos. The many people who allowed me to use their photos are thanked in the photo captions, but special thanks go to Rich Clark, in Indianapolis, and the wonderful people at the Brookfield Zoo and the Wildlife Conservation Society. Maili Poag did all of the beautiful line illustrations that begin each chapter. She also deserves special thanks, both for her talent and her kind consideration.

My editor, Julie Schroeder, did a brilliant job of hacking at my prose until it sounded as good as it possibly could under the circumstances. She is a genius.

Let me also take this opportunity to thank Clair Willcox, the acquisitions editor at the University of Missouri Press, for taking a chance on this book. I like to think he did it because he saw the enormous potential of my literary talent. On the other hand, it might have been because we were roommates in college.

My thanks to Judy Gagen and Caroline Lees for their encouragement and support and Cheri Asa, Steve Bircher, and Ron Goellner for letting me borrow their words so extensively. Ron and Steve had the added burden of having to work on Sundays, the day that I reserved for going to my office at the zoo and writing this book. They had to put up with my constant questions.

My wife, for her part, had to put up with two years' worth of one-day weekends. Even worse, she had to sit through countless readings of my drafts. In the end, if I had to thank one person, it would be her. Nobody else really had to put up with me to anywhere near the same extent, and nobody else helped me nearly as much!

Thanks to all of you.

Sailing with Noah

Introduction

A life without stories would be no life at all. And stories bound us, did they not, one to another, the living to the dead, people to animals, people to the land?

ALEXANDER McCALL SMITH

I often say that I don't remember a thing that I learned in first grade. I must have learned something, because I know the alphabet and I can count. Still, I don't actually remember anything with clarity. Most people at least remember their teacher's name or, better yet, some intriguing and wonderful lesson that seemed to stick with them over the years, but from the classroom, I remember nothing. I do, however, remember going on a field trip to Belle Isle Aquarium, now run by the Detroit Zoo. On my school field trip, I remember seeing a man in chest waders and rubber gloves who picked up an electric eel and placed it in a metal holder attached to a metal grid with light bulbs on it. The eel discharged, and the electricity from its body lit the bulbs.

I remember that like it was yesterday.

We call those types of experiences "landmark learning experiences." We know that they're important to us, but we don't really understand why. Perhaps, at that moment, I was astounded by the unique power of this living creature, perhaps I was amazed that something organic could power an electrical device, perhaps I was impressed by the fact that another human being was willing to grapple with such a dangerous creature, or perhaps I was just awed by the life force so tangibly manifest. Whatever it was, I changed on that day. I was never the same again.

Now I am the president of the Saint Louis Zoo. Before I got this job, I was the president of the Indianapolis Zoo and White River Gardens, at that time the only institution in America that was accredited as a garden, a zoo, and an aquarium. I cannot say that, because of my experience with the eel, I eventually came to work in a zoo that featured an aquarium. I could just as easily have become an electrical engineer or an eel monger. Either would have made just as much sense on the face of it. But I do often evoke that early childhood memory to help people understand why I do what I do for a living. More importantly, I tell people about it because it helps them to understand the influence that places like zoos can have on people.

Zoos certainly hold a great power over me.

I came to this work completely by accident. I was always more interested in school field trips than I was in school, but I did have a strong interest in science and did very well in the St. Louis Regional Science Fair. For college, I had my choice of a number of scholarships in the St. Louis area, but I wound up going to the University of Missouri in Columbia. I went there because I got to tour the campus on one of those miraculous spring days where winter becomes summer almost overnight (at least until spring reasserts itself with the more typical rain and cold). On that particular day it seemed that every single coed on the campus was out wearing a halter-top and skimpy shorts. I remember that day vividly, too.

While in college, I fell in love with the discipline of anthropology and decided I would become an anthropologist. I think I had the model of Indiana Jones firmly in mind, even though the movie hadn't come out yet. The author of my Anthro 101 text was a man named Marvin Harris, who taught at Columbia University, in New York City. That's where I decided to go for graduate school. I didn't know it at the time, but the anthropology department was then in its heyday as arguably the finest in the world. Margaret Mead was still on the faculty, as were Morton Fried, Conrad Arensberg, Ralph Soleki, and any number of other professors who stand among anthropology's elite minds of the twentieth century. It was an incredibly vibrant and thrilling era.

My only problem was that Columbia had not seen fit to raise its fellowship support levels since Eisenhower had been president. (That's president of Columbia University, not the United States, so it's even worse than you might think.) As a consequence, I had to have a job. Fortunately, Dr. Mead offered a pretty good idea. She had a joint appointment at the American Museum of Natural History and suggested that it would be good experience for any student to work in the anthropology department of the museum. To cut a long story short, I went to work at the American Museum for the Curator of North American Indians, Dr. Stanley Freed.

Dr. Freed's real interest was not American Indians. He had done fieldwork with American Indians early in his career, but his real love was India. He shared with me his passion for the country and people of India and, eventually, I went there myself to do my dissertation research.

After graduating from Columbia with a Ph.D. in anthropology, I went to work at the University of Michigan as a professor. Much to my dismay, I hated it. That's not completely true. I loved teaching, I thought the students were wonderful, and I genuinely enjoyed the people I worked with. I never felt, though, that I was really making a difference as a university professor. My area of specialization in anthropology is a discipline called cultural ecology—the study of the interaction of plants, people, and animals. My research in India had focused on land reform, showing how this reform had changed the ecology and, in turn, the very lives of the villagers of northern India. But I never got the same satisfaction from teaching as I did from "doing."

The real problem was that, frankly, there isn't much for anthropologists to do, aside from training more anthropologists, and I couldn't think of anything else that a Ph.D. in anthropology had prepared me to do. I wound up taking my problem to a psychiatrist on the assumption that there had to be something wrong with me. That was another one of those pivotal moments that I remember perfectly. I actually don't recall what we chatted about for the first twenty minutes or so, but I do have a vivid recollection of him asking me, seemingly out of nowhere, when I had been happiest in my life. "Oh, that's easy," I said without thinking. "That would be when I was a graduate student working at the American Museum." "Well," he replied with dead earnestness, "either go back to graduate school or go work in a museum. That'll be fifty dollars." He must have seen the light bulb come on.

Obviously, I wasn't going back to graduate school. But a museum! Why didn't I think of that? About three months later, I was on my way to St. Louis to work at the St. Louis Museum of Science and Natural History. Like most of the folks who work in natural history museums, I would spend a good deal of time keeping dead things dead, but it would prove important to my future in zoos. There is more than a little similarity between zoos, gardens, and museums. They all have collections, they have research and education programs,

they create exhibitions for the general public, and they are internally struc-
tured along much the same lines.

I spent ten delightful years in St. Louis, creating the exhibits and educational programs for a wonderful new institution. We started as a sleepy museum of science and natural history located in a beautiful park just outside
the city limits, and when we were done, we had built a shiny new science
museum, the St. Louis Science Center, in St. Louis's venerable grande dame
of Forest Park.

Forest Park is slightly larger than Central Park and is home to many of the
city's most prominent cultural institutions. These include the Muny theater,
the Missouri Historical Society, the Saint Louis Art Museum, the Science Center,
and, of course, the Saint Louis Zoo. It is a beautiful park that is esteemed by
everyone, not just for its cultural institutions, which are free to the public
due to the strong base of tax support, but also for its ambiance and history.
The park was the site of the 1904 World's Fair, an event that St. Louisans
talk about as though it happened just a few years ago. It was also the site of
the first modern Olympics held in the United States, in the same year.

During the entire ten years I worked at the Science Center, I was a regular
visitor to the zoo. Back then it was socially acceptable to have a beer at lunch
(a custom that, sadly, is no longer in vogue) and I would often spend my
lunch hour at the zoo, eating a hot dog and drinking a very large Budweiser.
I never dreamed that I would someday work there.

Shortly after the last phase of the new science museum was completed, I
received a call from an executive recruiter, or "headhunter." He wanted to
know if I had ever considered working at a zoo. "Sure," I said. (At least, I think
that's what I said. It's funny how we can remember moments of truth far better than, at least, the small moments of deceit.) The position the headhunter
had in mind was president of the Indianapolis Zoo.

Since I knew nothing about zoos, apart from what could be learned peer-
ing over a glass of beer on my lunch hour, I thought it would make a good
deal of sense to visit the Indianapolis Zoo before I talked to anyone about
working there. A week after the headhunter's call, I was on a plane to Indianapolis, armed with the some of the zoo's promotional materials and a map of
the city. I arrived early in the morning, before the zoo opened, rented a car,
and drove down Washington Avenue from the airport to the zoo. Since I had
plenty of time to kill, I stopped at a pancake house just past the airport for
breakfast.

Now, normally I hate when people say, "I swear to God this is true." But I have to say it now, because what I'm about to tell you is singularly improbable. Smiley's Pancake House was a run-down and eclectically decorated establishment. The decorations ran from reproductions of Red Skelton's clowns to (and this is really true) Elvis on black velvet. My booth seemed, at first glance, to be relatively unadorned. In fact, all it had was a framed quotation, centered on the wall over the table. The quotation was from a marvelous book, *The Outermost House,* by Henry Beston, and when I read it, I thought it had been placed above that booth specifically for me. The quote reads as follows:

> We need another and a wiser and perhaps a more mystical concept of animals. Remote from universal nature, and living by complicated artifice, man in civilization surveys the creature through the glass of his knowledge and sees thereby the feather magnified and the whole image in distortion. We patronize them for their incompleteness, for their tragic fate of having taken a form so far below ourselves. And therein we err, and greatly err. For the animal shall not be measured by man. In a world older and more complete than ours they move finished and complete, gifted with extensions of the senses we have lost or never attained, living by voices we shall never hear. They are not brethren, they are not underlings; they are other nations, caught with ourselves in the net of life and time, fellow prisoners of the splendour and travail of the earth.

Why they had this quotation over my bench in a roadside pancake house, I'll never know. But it is as beautiful a thought as I have ever seen expressed. I had never heard of Henry Beston before that day, but I have since read and reread his book, describing his yearlong stay on a then uninhabited New England beach. During my time in Indianapolis, I occasionally stopped by Smiley's and always checked to see if the quotation was still there. Then, about a week before I was scheduled to leave for my new job in St. Louis, I went back one last time, thinking I would offer to buy the framed quote to take with me as a sort of talisman. But Smiley's had morphed into something called the Red Apple Café, and all of the old decorations were gone.

But back to my reconnaissance trip to Indianapolis. After breakfast I drove through a light mist to the Indianapolis Zoo. It's funny how much the weather affects our perceptions of what is around us. Just as all of my memories of my years at the University of Missouri are colored by my first view of the campus on a spectacular spring day, all of my perceptions of the Indianapolis Zoo are tinged by the gray skies of an early winter morning. The grass was still a vivid green, however, and I had the entire zoo virtually to myself. The animals, thanks perhaps to the fine rain, were as animated as they might ever be, and I was thrilled to be there. I thought it was simply the most beautiful, the most real, and the most natural zoo I had ever seen.

The new Indianapolis Zoo was, at that time, about five years old. It was designed following a first-ever symposium on how to build a zoo, a symposium that was attended by zoo specialists, aquarium directors, and wildlife authorities from around the world. Their first conclusion was that animals should not necessarily be displayed taxonomically. Up until that time, the primates were displayed together in one location, the big cats in another, and so on. In fact, I grew up thinking that lions and tigers lived in the same places in the wild, in all probability because I always saw them next to one another in zoos. They decided instead to display animals according to the habitats they are found in. They suggested that the zoo be organized according to those habitats, and the Indianapolis Zoo was designed with major areas devoted to deserts, plains, forests, and aquatic environments. This meant that Siberian tigers, for example, were displayed in the Asian forest area, while African lions were displayed in the African plains area.

They also decided that the zoo should be essentially cageless, without bars or visible barriers between the animals and the visitors. Finally, they decided that, wherever possible, the exhibits should be multispecies in nature. This was thought to make for a much more exciting display, but obviously there are limits. Putting lions in with the Thompson's gazelles, for example, is a bad idea. However, this idea was applied in almost every major enclosure, and where the animals couldn't actually be together, the exhibits were arranged so that, from certain vantage points, it really did look like a pride of lions could at any second stir themselves for a run at the zebra.

When I first saw the zoo, in November 1991, it had matured considerably but still looked pretty new. The zoo had been built in White River State Park, an urban park that is located adjacent to downtown Indianapolis. The name had conjured images of old-growth forests, and my experiences in the Saint Louis Zoo, which has been in more or less the same location of Forest Park for a hundred years now, led me to associate zoos with beautiful old trees and shady paths. But White River State Park had actually been a light industrial area prior to the time the state acquired the land that the new zoo was built on. In fact, there were only three trees on the entire zoo property that predated the zoo's construction.

The good news, on the other hand, was that the plantings for each area were as appropriate as Indiana's climate would allow. The Asian forest, for example, had beautiful stands of bamboo lining the walkways, while the North American forests area had a mix of larger trees and beautiful understory plantings featuring dogwoods and other flowering trees. The Plains area featured a marvelous collection of grasses, and even though acacias would not grow in Indiana, they had managed to prune honey locust trees to make them look as though they had been magically transported from the Serengeti. Walking through the zoo on that morning, I thought the plantings *made* the zoo. Even though I knew absolutely nothing about the place, what challenges it faced, what problems it had, I knew within minutes that I wanted to be the zoo's president more than I had ever wanted any other job in my life. The fact that I knew nothing about zoos apparently did not faze me one bit.

I looked forward to working at the zoo from the moment I first walked through the grounds. Since that day, I have spent a good deal of time wondering how it is that we fall in love with something (or someone) before we really learn anything about it (or her or him). If you stop and think about it, this runs counter to the fundamental theory that underlies everything that happens in our educational system from first grade on. In every learning environment, including zoos, we base our approach on the notion that first we get information, then we gain awareness, then we gain interest, and finally we develop a passion.

For example, in almost every zoo in the nation, somewhere in their publications, exhibitions, or programming, you will encounter this quotation: "In the end, we will conserve only what we love, we will love only what we understand, we will understand only what we are taught." In many ways this quotation embodies the entire paradigm of conservation education, particularly as it is practiced in zoos and gardens.

The idea is relatively simple. It says, "Information leads to awareness, which leads to interest, which leads to caring, concern, and action." To put it in its most basic form, we believe that if we can get into people's heads, we can change their hearts.

But unlike the strategy of my first grade teacher, maybe our most important goal *shouldn't* be to impart information. Perhaps what we should be doing is trying to change attitudes and emotions. If we can get people to experience wonder or awe, concern or distress, or even love or hate, they will most surely be spurred on to learn more and act differently. In other words, instead of saying, "If we can get into people's heads, we can change their hearts," we should probably be saying, "If we can get into people's hearts, we can change their heads."

I have come to believe that the best way to ensure that people get the message is, first, to get them to care passionately about what you have to say. For example, in order to get people to care about the plight of gorillas in Africa, we must first get them to care about the particular gorillas they are looking at. The best way to do that, at least from an anthropologist's point of view, is to tell that particular animal's story.

Storytelling has been with us for as long as humans have been humans. In a way, storytelling combines information and emotion into one powerful medium. The best stories certainly impart information, but they also compel the reader's or the listener's interest. They don't teach so much as they allow insight. We draw our own individual conclusions from stories, but the best stories are the ones that make everyone care.

Zoos are much more than places that display animals; they are centers for education, conservation, and science. In fact, as you'll read later in this book, the Saint Louis Zoo has staff all over the world doing fieldwork (what we call "*in situ* research") and many staff at the zoo are devoted full-time to research efforts focused on the animals in our care (or "*ex situ* research"). Our research and conservation efforts here and abroad are vitally important, but the longer I work in zoos, the more I have come to believe that connecting visitors with the animals they see right in front of them is still the best service we can render. In other words, the animals that reside in zoos become ambassadors for their kind. They represent, as Henry Beston so eloquently states, other nations. By their very presence, they are impossible to ignore.

But animals cannot speak to us; they have voices we will never hear. We must speak for them, and we must speak in a voice much louder than we have used in the past.

We know what they would say if they *could* speak to us—they would tell us that time is running out. According to E. O. Wilson, the noted Harvard sociobiologist, fifty thousand species disappear from our planet every year—roughly six species per hour. Even the most conservative estimate of the number of species we lose per year would force us to conclude that we are undergoing the most pervasive loss of life-form diversity in the history of the earth. If animals could speak to us, they would also tell us that the world-wide conversion of wildlands to farmlands and developments condemns uncounted millions of wild animals each year to starvation, suffering, and, ultimately, to death. They would tell us that even our preserves are besieged and that, for some, there is no place of safety and sanctuary. They would tell us that if the mighty among them—the mountain gorillas, the Siberian tigers, or the Indian rhinos—cannot survive, then there is little hope for the meekest.

And they would tell us the stories of their lives—their triumphs and failures, their hopes and fears, and their joys and disappointments. That's why I decided to write this book. It seemed to me like a wonderful way to tell the stories behind the animals, plants, and people that call these zoos home.

The stories in this book are not arranged in any particular order. They jump between zoos in different cities and between countries on different continents. They jump between illness and health and between death and life. They range from fun and lighthearted to deeply tragic. If I picked them for any reason, it was because they are the stories that taught me about modern zoos and what they can (or cannot) accomplish. I became the president of a zoo largely ignorant of what zoos are all about. These are the stories that taught me the most. I hope that you enjoy them.

1

Escargot, Anyone?

One warm Sunday in May, I found myself standing on the balcony of my office overlooking the Saint Louis Zoo, watching the birds flock at the feeder below. A small house sparrow was under attack by a starling, which pecked viciously at the little bird. The sparrow would escape briefly, flutter a few feet, and the starling would once again renew the onslaught. A small ring of sparrows gathered around their tormented fellow, but they were obviously incapable of stopping the larger bird. A flock of pigeons, oblivious to the commotion, continued to feed only a few feet away from what appeared to be a death struggle.

I was tempted to go run down the steps and stop the starling, but then I thought otherwise. First, it occurred to me that this was a part of nature. I had no right to interfere in the lives of those birds. Then it occurred to me that I probably *did* have a right to interfere. It was my feeder, after all, since I was ultimately responsible for having it there and keeping it full. But, before I could move, I had a third thought. Not a single one of those birds were supposed to be there in the first place. All three species—house sparrows, starlings, and pigeons—are interlopers. They were not present in North America until humans brought them here and released them. Many would argue that, in a perfect world, the North American populations would all be exterminated. In fact, that is the core of one of the fundamental debates in conservation. The question is, should we attempt to undo the damage caused by man and restore habitats to the forms they took before humans dramatically altered them, or are changes like the introduction of totally new species just part of the normal course of things, whether that introduction is at the hand of man or the hand of nature?

By the time I was on to the third thought, the little sparrow regrouped, mustered its strength, and flew to a high tree some fifty yards away. The starling remained in hot pursuit. I never saw if the starling continued the attack or gave up, but by that time, I was too busy mulling over the larger questions of environmental ethics to worry about the outcome of the struggle. I was also a little relieved to have the whole issue taken out of my hands.

House sparrows, pigeons, and starlings were all deliberately introduced into North America, each for a very different reason. Of the three, the story of the introduction of the starling is probably the most bizarre. On March 6, 1890, a man named Eugene Schieffelin walked out to Central Park in New York City with eighty European starlings in small wooden crates and, quite deliberately, changed the ecology of North America.

Schieffelin, as it happens, was a huge fan of William Shakespeare. His life's dream was that all of the birds mentioned in Shakespeare's work might someday come to inhabit his city. No one knows exactly why he thought this would be of benefit to the populace at large. Perhaps he thought that the city would be a better place if Shakespeare's larks and nightingales were heard singing in the early evening as New Yorkers strolled through their beautiful park. Maybe the starlings were just an afterthought. After all, they're only mentioned in one place in all of Shakespeare's works—*Henry IV, Part I*. He may have included them out of a sense of completeness, while he thought mostly of introducing birds with sweeter songs.

We'll never know, but we do know that a single flock of starlings (called a murmuration) can grow up to a million or more birds. They can blanket the sky and, when they descend on a farmer's crop, they are ruinous. They plague our cities, covering the ground beneath their roosts thick with excrement. They torment our native birds, robbing nests and killing the young. From the original eighty birds he released in 1890, along with another forty he added in April 1891, Schieffelin is responsible for the presence of the two hundred million starlings found in North America today. Most people, if they thought about it at all, would wish that there weren't any.

Pigeons have been in North America much longer than starlings. In 1604 the great French navigator Champlain piloted a boatload of colonists to the New World. In exchange for a guaranteed monopoly on the fur trade, Champlain and his patron had agreed to establish a settlement. Their first stab at empire building was a small village near the mouth of the St. Croix River, near the boundary of what was to become Maine and Canada. It nearly ended in disaster, with the little colony barely making it through the first winter. They elected to move to what is now Nova Scotia, where they fared little better. After a second winter they gave up, but as they sailed out to sea, they encountered a much larger ship bringing reinforcements. With a fresh load of supplies, they turned around and tried a third time. It proved to be the charm.

Among the supplies they received were pigeons. In her incredible book *Tinkering with Eden,* Kim Todd devotes a whole chapter to the introduction of pigeons in North America. At the time, pigeons were more than just a source of food in Europe. The nobility kept huge flocks in giant pigeon coops, called dovecotes. It was a basic right for the nobility, and it came along with the right to own the land that the peasants tilled. Imagine thousands of birds flying over a peasant's fields, eating their grain before they could harvest. One of the first things that angry peasants did at the onset of the French Revolution was to release the pigeons, a living symbol of the tyranny of the elite.

On the other hand, they were good to eat and fun to breed. Over time, their role as a source of succulent dinner fowl was usurped by an interest in breeding pigeons as a hobby. By the 1800s people were breeding pigeons with a variety of features that fairly boggle the mind. According to Todd,

> Birds were bred for plumage that curled out rather than in, short beaks, long legs, necks without feathers, and good posture. Many varieties looked like ordinary pigeons viewed in a fun house mirror. Tumblers flipped over backward in the air in a flurry of feathers, an excellent parlor trick. Pouters' chests swelled so high that breast feathers often buried their faces. Fantails displayed prominent tail feathers bristling in an erect semicircle, like a peacock's. Runts, bred for size, boasted wingspans of three feet.

They also bred the birds for navigational skills and released homing pigeons in competitions.

By the early 1900s, the pigeons, beneficiaries of three centuries of selective breeding by humans, were ready to launch a takeover. Their target was urban America. The present-day urban pigeon has elements of many of its selectively bred ancestors, many of whom were released or escaped from their coops to join their brethren. While many of the traits that hobbyists had bred for were singularly unhelpful, some, like navigational skills, conferred a selective advantage on pigeons. The trait that centuries of human interference has produced that probably helps the pigeon most of all is that pigeons have been bred to get along well with humans. They're pretty much the perfect urban bird, even though they started their relationship with us humans in the farmer's field.

Sparrows, like pigeons, have a relationship with humans that is based largely on agriculture, but their history in the New World is nowhere near as long. The introduction of house sparrows occurred in Brooklyn in the spring of 1851. A man named Nicholas Pike, a member of the Brooklyn Institute, was appointed as chairman of a committee devoted to sparrow introduction. The first eight birds they released died, but their subsequent efforts, in 1852 and 1853, were an unqualified success. From Brooklyn they radiated out through the eastern United States and into Canada. More releases occurred later in the western United States, in places like San Francisco and Salt Lake City in the mid-1870s. The largest single release was undertaken by the city of Philadelphia, where the city government imported one thousand birds for release.

I have heard that the birds were originally introduced specifically to eat the canker worms that, in turn, were eating Brooklyn's shade trees. While this may be true, it was also widely believed that sparrows were a truly "beneficial" bird. They were thought to thrive on a steady diet of agricultural pests, eating the insects that eat our crops—but this is not exactly true. It is estimated that most (60 percent) of their diet in rural areas actually comes from eating the crops they are supposed to protect. But they do "good," by our point of view, in that another 36 percent of their diet comes from eating weed seeds, and the remaining 4 percent comes from eating insects. So, while they may be helpful, on balance they eat more crops than they eat pests.

The house sparrow thrives in association with humans. House sparrows are not often found in forests or open grasslands, but you can always find them wherever you find people. Their "native" range is from Britain all the way to Siberia, ranging down into northern Africa and across to Arabia, India, and all the way to Burma. Thanks to humans, though, they can be found in South America, southern Africa, Australia, and New Zealand as well as, of course, here in America. And they are found here in great numbers. The population in North America alone is estimated to be some 150 million. While you can argue that each bird doesn't eat much, you have to agree that "not much" times 150 million is a whale of a lot. House sparrows are an environmental disaster of the first order. Furthermore, they're an environmental disaster in most of the world.

But they're *our* environmental disaster. They coevolved with people. Much like the common rat, their story, and ultimately their fate, is linked with our own. They go where people go in general and where agriculture goes in particular. When they were first introduced in New York they ate the feed and waste of horses. Horses were, of course, everywhere in the 1850s. There was no shortage of food, even in an urban area. Now their numbers have begun to decline in the eastern and central United States as our agricultural practices continue to change. I wouldn't bet against the house sparrow, though. They've been around a long time.

Most ecologists are profoundly aware that agriculture itself changes everything. The existing habitat is stripped absolutely bare of its original plant base. With the demise of the plants go the animals that are dependent upon those plants. But agriculture does not just transform the local environment. Agriculture, ultimately, transforms our world and everything on it.

Anthropologists recognize three great ages that mark the transcendence of human beings on our planet. The first is the Paleolithic, the "old stone age." This is the period when humans first began to shape their world using tools, mostly primitive hand axes made from big hunks of rock. While not very elaborate, these tools gave us a great advantage over the species that constituted our competition in the race for survival. The Mesolithic, the "middle stone age," is marked by the development of far more sophisticated stone tools that are beautifully crafted and that feature fine, razor-sharp edges. The third era is the Neolithic, or "new stone age." This era is marked not so much by the quality of the tools produced, but instead it is marked by what the tools were used for. They are used exclusively for agriculture, the heart of the most important revolution our species has ever seen.

We know that agriculture got its start in the Near East. To this day, wild wheat grows in Turkey, and anthropologists estimate that with no more than a flint-bladed sickle a family could harvest enough wild wheat in three weeks to last a whole year. Grain meant life, but it also meant a change in lifestyle. No more would small groups of hunters roam the planet in search of game, killing what they could and gathering enough to keep their little bands fed from one day to the next. Instead, people would settle in villages where they could tend to their crops. The plants themselves would change. Wild grasses, the precursors to our barley and wheat, were naturally adapted to reseed themselves. Farmers, on the other hand, want the seed to stay on the stalk so that it can be easily transported to the threshing floor, and then cleaned, husked, winnowed, and ground into flour. Gradually, plants that were more suited to agriculture were developed, a process that continues to this day at companies such as Monsanto, the agribusiness giant, which is headquartered just down the road from my office.

Along with plants, humans domesticated animals. Sheep, goats, and cattle soon became a regular feature of the first villages. Wild herbivores were captured and then selectively bred to make them more docile, better milk producers, better meat producers, and so on. At some point the originally wild plants and animals became true domesticated plants and animals, many of which, plant and animal alike, can no longer survive without the aid of humans. Other plants and animals (like our sparrow), although never actually becoming domesticated, changed over time to coexist with the new way that people lived on the land.

We could, if pressed, survive without the house sparrow. In fact, many of us can cheerfully imagine a United States without them. But we could not survive without domesticated plants and animals, at least not in the style to which we have become accustomed. Agriculture changed *Homo sapiens* from a rare to an abundant species. We ceased to be nomadic and settled down in villages. Thanks to agriculture, a handful of people could feed many. (In fact, today about 3 percent of our population feeds our entire nation.) Once we no longer had to roam restlessly over the earth in a ceaseless quest for food, we were free to develop a host of other new inventions that would, in turn, change our lives over and over again. By 6000 BP (years before the present), humans had invented spinning and weaving, ceramics, the smelting of bronze, bricks, pottery, sailing ships, wheeled vehicles, writing, the calendar, and a system of weights and measures. That, of course, was only the beginning. Soon our capacity for invention would yield a list that would be much longer than this entire book. The list of social changes is just as important. Some 6000 years ago, according to Marvin Harris, "for the first time, human communities became divided into rulers and ruled, rich and poor, literate and illiterate, townsmen and peasants, artists, warriors, priests, and kings."

In a classic article, "The Spearman and the Archer," an anthropologist named Alice Brues argues that the Neolithic Revolution changed our bodies along with our lives. She begins by arguing that, as the technology of weaponry changed, so did our bodies. For example, we know from skeletal remains that those guys wielding the clubs with large rocks attached to the end, typical of the Paleolithic, had a short, powerful physique ideal for swinging a heavy implement at close quarters. But spears are more effective than clubs. Put a razor-sharp spearhead on a stick and hurl it from a distance and you'll be a more successful hunter and, in a contest with a guy with a club, a more successful fighter. Where you find spears, you find long, thin people, and from a physicist's point of view, that makes perfect sense. But the arrow trumps the spear in the same way that the spear trumps a club. When I was growing up, we all wanted, whether we knew it or not, to have the body of an English archer. The perfect body type for an archer is a powerful, V-shaped upper body with strong muscles through the chest and arms. To this day, when I stand in front of the mirror and imagine that I look other than like a paunchy, fiftyish, zoo administrator, I imagine myself looking like an archer.

The most powerful weapon of all, though, is the hoe. It hardly matters what technology of weaponry you have if you're up against a society that has a hoe. As we've seen, hoes mean agriculture, and agriculture means the ability of the few to feed the many. Once you have agriculture, you can have full-time warriors. They can travel long distances and bring everything they need to sustain a campaign along with them—they can, and they did. Those early villages eventually became empires, empires that hungered for more lands to conquer. More land meant more power, more security, and greater success. It also meant ecological change on an almost unimaginable scale.

This was the true beginning of empire building, and it was built on the sturdy backs, the short, powerful legs, and the strong arms of the peasant. This body type, in the ultimate version of coming full circle, was basically the same as that of the first primitive wielders of the club.

As those early empires expanded, first in Mesopotamia and later in places like Egypt, around the Mediterranean, China, the New World and so on, the natural landscape was altered forever. We cannot reclaim it or restore it. The best we can hope to do is save or restore small fragments of it. Here in the United States our vast prairies are gone. The eastern forests no longer stretch unbroken across the landscape. The Great Plains are largely under the plow, and the vast herds of bison that once roamed the plains are largely absent.

And so we have answered, at least in part, the first big question that conservationists ask: "Should we attempt to undo the damage caused by man and restore habitats to the forms they took before humans began to alter them?" The answer must be, "No, because it's simply impossible." The scale of environmental change is entirely too vast to even contemplate. Our planet is changed forever. It cannot be changed back.

However, what about the small, remaining fragments that are representative of what the environment was like before the great Neolithic Revolution? Humans have released cats, goats, and rats in the Galapagos Islands. Shouldn't they be exterminated before they can destroy the unique wildlife of that isolated and magnificent island chain? What about exterminating mountain goats in the Olympic Peninsula of Washington? What difference does it make if humans brought those animals to those two very different, but very pristine environments? The fact is, they're there. Don't they have a right to stay there?

The mountain goat really is a classic case in point. Lee Lyman tells the story in his book *White Goats, White Lies,* but here it is in a nutshell: Franklin Roosevelt established the nine-hundred-thousand-acre Olympic National Park in 1940. The goats had been in the park, at that point, some fifteen years, having been released with the permission of the US Forest Service and the Washington State Department of Fish, Game, and Furs. When the area became a national park, it fell under the jurisdiction of the National Park Service. The NPS took a very different view of the mountain goats than did the Forest Service. They wanted the goats out, partly because that's their policy. It's a very clear-cut mandate; it reads, in part: "Management of populations of exotic plant and animal species, up to and including eradication, will be undertaken whenever such species threaten park resources."

Mountain goats, like their domesticated cousins, will pretty much eat anything that grows, and that part of the Olympic Mountains has some plants that are very rare and exceedingly delicate. It seems pretty clear that the goats were likely to eat some of those rare plants into an untimely oblivion. When the Park Service announced that they would begin to cull the goats, animal rights groups had a field day. They were certain that the goats' right to continue to live in the park outweighed the rights of the plants to continue to live at all.

I believe that we do have an obligation to eradicate species introduced by humans into a relatively pristine habitat if and when the introduced species threatens to cause the extinction of the plants or animals that lived there before we began our meddling. Where we cannot save the threatened species *in situ,* that is, in the wild, then we have a special obligation to ensure their survival in places like zoos. We have an even more special obligation to try to return them to the wild.

Sometimes that proves impossible. Take the case of the partula snail. The family Partulidae is a group of tree snails that evolved in the tropical South Pacific Islands. There are three genera in the family; one of these is *Partula,* which is far and away the most diverse, with over one hundred different species. The range of *Partula* is vast, covering some five thousand miles from

Palau to the Society Islands, but it is on the Society Islands that *Partula* took off. The Society Islands have a variety of different microenvironments that range from moist, tropical valleys to dry, windswept ridgetops. Where one finds pronounced environmental variation, there is usually a lot of speciation. In other words, animals adapt to different environments and gradually evolve into totally different species. In fact, this small group of islands has about fifty different species of *Partula*.

From a human's point of view, partulas are relatively harmless. They eat microscopic plants that grow on the undersides of leaves, but they don't eat the leaves themselves. They're attractive little guys that vary in color from pale to dark brown, and many of the species have a lovely spiral pattern on their shells. Even the biggest species of *Partula* only get to be about an inch long.

Partula are very well known to science. They constituted a major part of the battleground between the proponents of Lamarck's theory of evolution via acquired traits and the proponents of Darwin's theory, which was based on natural selection. Both sides used the evidence provided by the speciation of *Partula* to support their theories. The principal "biographer" of the genus was a man named Henry Edward Crampton, who studied them on Tahiti and on the neighboring islands for fifty years. Crampton's work inspired others and, in recent times, three different scientists have continued to build on Crampton's monumental work. The scientists, Brian Clarke, Jim Murray, and Mike Johnson, did most of their work not on Tahiti but on the nearby island of Moorea, some fifteen miles away.

Moorea is better known by its stage name, Bali Ha'i, from the musical *South Pacific*. When the three scientists began their work on Moorea in the 1960s, they thought they would be studying how the snails evolved into new species. They never dreamed that they would wind up documenting their demise.

The problem began in 1967, when a large African tree snail, *Achatina*, was deliberately introduced on Tahiti as a source of food. When I say large, I mean large—they can grow up to nine inches in size, and they're delicious, too, or so I'm told. Within a short period of time, though, *Achatina* outwore its welcome. Unlike *Partula*, *Achatina* eat leaves. The huge snails began to strip the island's crops and gardens bare. It wasn't long before they hopped a boat to Moorea, where their population continued to explode. Our three scientists write of removing two wheelbarrow loads of the giant African terror from the *inside* of one house. Imagine how many were *outside*.

Unlike the mountain goats, where we debate endlessly their right to trample the rare wild plants of the Olympic Mountains, there was no debate about the need to do something about the *Achatina* problem. When wild plants are threatened, it's one thing, but when agriculture is at risk, it's entirely another.

At the peak of the crisis, the French government of Tahiti turned to a snail named *Euglandina rosea* for help. *E. rosea* hails from Florida and is a carnivorous snail—it eats other snails. I don't know how it eats other snails, so don't ask. Honestly though, can you imagine the thrill of the hunt, as *E. rosea* "races" toward its intended victim? The horrible thudding sound as it crashes into its prey, crawling at full speed? The "lightning quickness" of the kill?

Clarke, Murray, and Johnson all thought that the idea of releasing *E. rosea* was, well, dumb. In fact, most ecologists protested the government's decision and, as it turns out, they were right to object.

E. rosea never developed much of a taste for the giant African snail. Maybe it was just too big to kill and eat. After all, *E. rosea* itself only grows to a size of two and a half inches. In any case, *E. rosea did* develop a hearty appetite for *Partula*. *E. rosea* was introduced on Moorea in March 1977. By 1984 it had eaten every last member of one of the seven species of *Partula* found on that particular island, and by 1988, it had eaten every single one of the remaining six species found on the island. Scientists estimate that *Partula* have been around for 1.5 million years; it took a little over ten years for an introduced species to wipe them out.

Ah, but it turns out that they weren't gone entirely. Once the three scientists realized that the first species was extinct, they put out a call to save the remaining six species by breeding them in zoos. Their plea came to the attention of one of the most remarkable men the zoo world has ever known—Ulysses S. Seal. Ulie was a biochemist with a post-doctoral specialization in endocrinology. He worked as a cancer researcher at the VA Hospital in Minneapolis at about the same time that the city was involved in building a new zoo. Ulie got involved with the zoo, and the result was that he brought the idea of applying basic science to conservation. In fact, some would say that our modern, computer-based species-management programs began with the efforts of this charismatic, energetic man.

As luck would have it, Ulie was scheduled to speak here at the Saint Louis Zoo in April 1987 for a regional zoo conference. By all accounts, Ulie held his audience spellbound with his presentation on the partulas' plight and, shortly after that conference, in 1990, the zoo world created a coordinated breeding plan to save these invertebrate species—the lovely *Partula* snails. Ron Goellner, now the zoo's general curator, still coordinates this plan in America, as he has since a small group of zoos came together here at the Saint Louis Zoo. The US zoos first met in St. Louis in 1991 to plan the partulas' future and, fortunately for the snails, a core group of zoos from the United Kingdom (led by the Zoological Society of London) and the United States developed a management plan that works very well.

As always, the first two questions Ron had to answer were, "How do you feed them and how do you breed them?" Fortunately, their breeding habitats were pretty well studied. Ron describes them as follows:

> *Partula* are hermaphrodites—each individual bears organs of both sexes. During the courtship ritual, they go through a series of gestures, finally approach, and then wait an hour and a half before they breed. After this process, one animal is impregnated. If things have gone well and they have an affinity for each other, they will court again and both will leave pregnant.

I love the part about how they wait for an hour and a half. I mean, it's just perfect.

Feeding them, on the other hand, was another matter entirely. We don't have a ready source of microscopic plants that grow under the leaves of Tahitian trees. We first tried to feed them a diet of oatmeal and powered chalk (for calcium), but we have gradually developed a much more sophisticated recipe. If you ever have a partula snail over for dinner, all you need to do is blend the following ingredients with water, creating a sticky paste:

> 3 teaspoons rolled oats
> 3 teaspoons ground cuttlebone
> 3 teaspoons powdered nettle
> 1½ teaspoons trout chow
> ¼ teaspoon calcium/vitamin A powder
> 20 milligrams vitamin E powder

The work of feeding and caring for the snails would not fall to Ron, though. Instead, the zoo developed a small cadre of volunteers. The first five volunteers to work with *Partula* were Nancy McLean, Millie Dellinger, Trish Larsen, Ellen Miller, and Suzanne Sandridge. Between them, they cared for the partulas every day. According to Ellen, it was, and still is, a fun way to relax. She describes it as kind of like watching goldfish swim around in a bowl, only slower.

The husbandry routine is the same today as it was back in the early 1990s. The snails are checked every day and fed twice a week. Volunteers clean the snails' homes, feed them, and keep them moist, all the while tracking their reproduction and growth in journals. The volunteers' first task is to check if the snails are "sticking" to the sides of their homes. Sticking snails are happy snails.

They live in clear plastic sandwich boxes that are lined with moist paper towels and covered by perforated plastic wrap. After the volunteers finish their initial observations, they clean the boxes by gently lifting out the paper towel. To do this, they must remove each little snail and snail egg from the toweling. Occasionally they find an escapee crawling up their arms. After the little box is cleaned, a new moist towel is put on the bottom and their food is smeared on the side.

So the *Partula* live on, not in their beautiful island paradise, but in small, plastic sandwich boxes stacked here in the Saint Louis Zoo in the research area of the Monsanto Insectarium and stacked in the back rooms of other zoos around the world. Will they ever roam free on their island again? Probably not. Right now we have an experiment going on in Moorea that involves creating a predator-free zone. It's a sixty by sixty foot area surrounded by supposedly snail-proof barriers. It's labor intensive, though, and it turns out that *E. rosea* is pretty good at sneaking in. In all likelihood, the *Partulas*' future is here. They are safe, but not saved. They will not be saved unless and until they can return and breed on their island home. Perhaps it is not enough, but at least it is something. After all (to paraphrase Aldo Leopold), the first rule of intelligent tinkering is to save all the pieces!

James Michener based his mythical, idealized island of Bali Ha'i on Moorea, creating the story for one of the most beautiful musicals ever produced. In return we sent the island *E. rosea* from Florida. It hardly seems like a fair exchange. On the other hand, the African snails did almost get their revenge. In 1966 a boy smuggled three giant African snails into southern Florida, and his grandmother released them into her garden. Seven years later, more than eighteen thousand snails were recovered by the state of Florida. It cost the state a million dollars and took over ten years to get rid of them.

Escargot, anyone?

2

IronKids Bread

At first I didn't get the joke. It never occurred to me that there might be something funny about having five elephants with diarrhea. I might have been the first to laugh at a news story about a fire in a match factory or a flood in an Alka Seltzer factory, but I never suspected that the newspaper would poke some gentle fun at our sick elephants. I wasn't particularly angry about our elephants being the butt of their joke. (See what I mean? You can't help but smile when you read the word *butt* in a sentence about elephants.) . . . But our elephant keepers were infuriated.

Elephants are their life. Most keepers have several different species under their care, but elephant keepers are pretty much specialists, especially at the Indianapolis Zoo. They may rotate to another area briefly, perhaps to cover for another employee who is on vacation or is sick, but otherwise they spend eight hours a day working with elephants. They clean the barn when they arrive in the morning, then they feed the elephants, bathe them or, in the winter, rub them with oil, then exercise each elephant in turn, and, during the times when they are not providing for the elephants' basic needs, they train them. All day, every day, they work with elephants.

This is especially true at Indianapolis because they practice what is called "free contact" elephant management. There is a spectrum of elephant management techniques in the zoo world. At one end of the spectrum we have the "restricted contact" school of thought that says that while elephants are, for the most part, gentle creatures, they are big and clearly can be dangerous. Keepers should not be near them unless they are separated by bars or the elephant is in some sort of restraint mechanism (like a squeeze cage—a device that holds an elephant between a set of moveable bars and immobilizes the animal).

Indianapolis is toward the other end of the spectrum. They believe that elephants and their keepers should be in constant, close contact. They feel that elephants benefit from the mental stimulation and sustained exercise that characterizes a "hands-on" program of care and training and, while they acknowledge that there is some risk, they feel that the benefits far outweigh the potential problems. Now, I'm not suggesting that elephant keepers from zoos that don't practice free contact management aren't emotionally attached to their charges—they most definitely are—but it is clear to me that Indianapolis's elephant keepers have very strong emotional bonds with their elephants and that the elephants are very close to their keepers.

That begins to explain why they didn't find the "elephants with diarrhea" jokes very funny. They would have been somewhat offended by anyone making light of an animal they take so seriously. The real reason they were infuriated, however, is because our elephants were likely to die from the dehydration associated with diarrhea. It was not a laughing matter; it was a matter of life or death.

I had not been working as the new president of the Indianapolis Zoo for very long when I first learned that all five of our female African elephants were violently ill. I was sitting in my office when I got a call from Debbie Olson, the curator of the African plains area of the zoo. Debbie is soft-spoken and, like many of the people attracted to zoo work, she sometimes seems more comfortable around animals than around people. "We have two sick elephants," she said, "and it's pretty likely that all of them will be sick soon if it's what we think it is. The vet's almost certain that we've got an outbreak of salmonella."

Salmonella is an infectious bacterium, named after the American veterinarian Daniel Salmon, who first isolated it in 1885. Ten species have now been identified, and one of them, *S. typhimurium,* commonly causes what we refer to as food poisoning. Salmonella food poisoning can strike many different animals, including elephants and humans. In humans an attack can last three to seven days. Mild cases are treated with an antidiarrheal, while more severe cases might require antibiotics. It was pretty clear from the outset that our elephants had a severe case. Our elephants started showing symptoms on May 1, 1994. Only a few years earlier, an elephant at the Metro Toronto Zoo had died of salmonella. In that tragic case, the elephant died on exhibit in the bathing pool. It took three days to get the body out of the pool, all the while attended by a heartbroken staff.

The first two elephants to show symptoms at Indianapolis were Cita and Ivory. Cita, at first glance, appeared to be our second largest elephant. She was not as massive as Sophie, our matriarch, or as tall as Kubwa but, being more compact, she actually looked like she would weigh more than Kubwa. She didn't weigh more—she was just a beautiful, classically proportioned animal.

At first Cita didn't act at all sick, but the keepers noticed off-color feces with a distinct aroma. In retrospect, her appetite may have been down a little, but it certainly wasn't noticeable at the time. Almost as soon as Cita stopped eating, the illness spread to the rest. Ivory, our ten-year-old baby, stopped eating and, shortly after that, stopped drinking. She appeared listless and lethargic. Next came Kubwa, and eventually all five tested positive, although Sophie and Tombi never suffered the most violent symptoms.

Like the elephants in almost all North American zoos, the elephants at Indianapolis were not related. Each of them had come to the zoo in a different way and all of them had fascinating lives prior to their arrival. Kubwa was caught in the wilds of Africa when she was about two years old. From age two through age seven, she lived alone at Indianapolis's old zoo. This is no longer allowed to happen in American zoos. Elephants are social creatures and need to be around other elephants in order to develop normally. Kubwa, having lived alone for a fairly long time, was probably better around people than she was around elephants. With other elephants she seemed awkward and unsure of herself, particularly with older, larger, and more dominant elephants. Physically, she was not as graceful as other elephants. Her "ankles" (elephants walk, quite literally, on their tiptoes) were weak, and perhaps that contributed to her lack of agility. By the way, elephants are quite agile. All of the elephants in the Indianapolis Zoo could walk out on a six-inch balance beam and turn around in the center without slipping. I'd have a problem doing that.

When Kubwa was seven, the zoo purchased a companion elephant for her. (They had provided her with a goat as a companion, but that didn't last too long. Kubwa accidentally rolled over on the goat and, well, you can imagine the rest.) Kubwa clearly needed an elephant friend as well as a better enclosure. In the old zoo she was kept in a circular exhibit, and the public was allowed to get very close. She occasionally grabbed cameras from visitors, and on one occasion she ate a lady's purse. In fact, Kubwa still walked in tight circles if she got excited—doubtless as a result of the time spent in her smallish pen. In recognition of all this, the zoo acquired a wonderful baby elephant that we christened Ivory.

We acquired Ivory from a company called International Animal Exchange, and the process Debbie used in selecting Ivory out of a herd of about twenty other baby elephants shows how little we knew about elephants at that time. Ask yourself this, "What would I look for in picking a baby elephant?" Not having been in the situation before, Debbie assumed that one would pick a baby elephant in much the same way that one would pick a puppy. You look for a bright, active, outgoing, gregarious puppy—one that is personable and clearly a leader. Ivory would have been the perfect puppy, except that she's an elephant. The older, more dominant elephants do not necessarily esteem her leadership qualities, so the fact that she was exceptionally smart was a source of pride *and* aggravation (in roughly equal measure) to her keepers.

In retrospect, Debbie claims she would have probably picked a less dominant animal if she had it to do over again. Ivory was always a challenge. Quick to learn, she also used her intelligence to test and occasionally outwit the staff. She was physically quick, and that, too, presented challenges to the staff. Because of this, when they trained a new elephant handler, Ivory was always the last animal the trainer was allowed to work with.

Kubwa and Ivory grew to be very close. To this day, they still greet each other loudly if they have been separated for even a few minutes. But since Ivory was, for so long, our youngest elephant, every elephant seemed to like her. They would stand over her if she lay down to rest, and she seemed to be able to pilfer a little of the other elephants' hay whenever she wanted—all this despite the fact that she was then a "teenager" and almost as big as the others.

Our most colorful elephant was Cita, the first to contract the salmonella bacteria. She was acquired in 1988, the fall before we opened the new zoo, and had already lived a very interesting life prior to coming to Indianapolis. After leaving Africa, she took up residence in a drive-through safari park in California. She had a baby that was stillborn and, shortly after that, the park closed. Cita was sold to a company called Animal Actors of Hollywood, where she embarked on her second career.

Her first film, *Sheena* (as in, "Queen of the Jungle"), was shot on location in Africa, and Cita starred in it along with Tanya Roberts (of *Charlie's Angels* fame). Her second film was *Pee-Wee's Big Adventure* and, as in her third film,

The Color Purple, Cita is only on-screen for about ten seconds. Her career, while short, was still impressive.

After her stint with Animal Actors, Cita was acquired by private owners, who used her to give rides and rented her out for other purposes. We met her when her owners brought her to Indianapolis to give rides back in 1987. Everyone liked her immediately. Furthermore, at the zoo she finally found a stable, secure home and blossomed. Like Ivory, she tests the limits of her handlers constantly. She knows exactly how much she can get away with before she gets a verbal reprimand and seems to know just how far she can push each different staff member. Her keepers say that sometimes she responds in an atypical way. They suspect that she may have been subjected to some bad training techniques (by definition, techniques that use physical punishments instead of positive rewards) at some point in her life. For example, when we first got her, she wouldn't lie down and was very sensitive about her hindquarters. We suspect that someone may have used a cattle prod on her to get her to perform, but we'll never know for sure.

Our matriarch was Sophie. She was a huge elephant, weighing all of 9,600 pounds. She went to Canada from Africa at age four and, like Cita, she lived in a drive-through safari park. We were looking for another dependable elephant for visitors to ride and to go along with Cita; everyone we called seemed to think that Sophie would be the best. We paid forty thousand dollars for her—a record price for an African elephant in North America—but she really is a wonderful animal. Her first trainer, however, must have been short-tempered and perhaps abusive as well, because Sophie always seemed to be afraid that she would make a mistake. On the other hand, she was our most dependable elephant, and the staff felt a high degree of trust and faith in her. In fact, all the new keepers learned first with Sophie.

The sad thing is that, despite her anxiousness to please, it took Sophie longer than any other of our elephants to learn a new behavior, and there were some things we just couldn't do with her. For example, we did some research on elephant memory, and one of the tests required the elephants to make a choice on their own. Sophie couldn't bring herself to make a choice. She was too afraid of being wrong.

Because of her size, Sophie was the boss. If she wanted something from one of the other elephants, she would get it. And she'd get it without a challenge. She was not, however, a good, assertive leader. She didn't provide guidance to the other elephants or step in quickly to resolve disputes. She often allowed Cita to have her own way, and Cita had even, on occasion, struck

Sophie with her tusks. Perhaps this is the normal course of things, even in the wild. Leadership must evolve over time, and younger, stronger animals must eventually replace their elders.

The last elephant the zoo acquired was Tombi, who came to us in 1989 from a zoo in Michigan. Tombi was raised as a single elephant from shortly after birth and the zoo in Michigan could not provide optimal care for her. As I said, we no longer keep elephants alone in zoos as a matter of policy, and now, in fact, we are phasing out elephants in zoos that can only keep two animals at a time. Within a few years, all zoos that have elephants will have to have at least three. Tombi worked well for her two male handlers, but she chased and slapped at any woman who tried to work with her. We agreed to take her on loan, and her apparent problems with working for females never manifested themselves. Perhaps she sensed that the male handlers at her last zoo didn't want her to work with females.

In any event, Tombi was pretty easy-going and always willing to please. She preferred to be around people, and other elephants just scared her. After all, until her arrival in Indianapolis, she hadn't seen another elephant since just after her birth. At best, her memory would be poor and incomplete.

Tombi's problems with our other elephants started before she even got off the moving van that brought her down from Michigan. She heard the other elephants before she saw them, and her first reaction was to "get big." Elephants can flare out their ears, push out their chests, and lift their heads in a way that can only mean, in the immortal words of Michael Jackson, "I'm bad, I'm bad, you know it." She flared when she first saw the other elephants. What she should have done as the new girl (and a small new girl at that) was to lower her body, present her rump, and show respect. The other elephants quickly set about teaching poor Tombi some proper manners, but the damage was done. For years the other elephants continued to pick on her, but she made slow, steady progress toward learning the full range of proper elephant behavior.

Interestingly, Sophie, who was at the top of the dominance hierarchy, got along well with Tombi, as did Ivory, who was the baby and thus at the bottom of the hierarchy. Kubwa and Cita continued to harass Tombi. Perhaps it was a classic case of displacement—or perhaps they did it for their own amusement.

Despite all their behavioral differences, the five elephants reacted in much the same way after they were stricken with salmonella. All five became listless. They appeared, even to the untrained eye, very, very tired, and would often just lean against the wall of their stalls. Their eyes seemed dull and flat, and Cita, Kubwa, and Ivory stopped eating and drinking entirely.

Elephants are very poor at digesting food. As a consequence, they must eat an enormous amount of food every day. Watching them in the wild, you will see them walking and eating nonstop, unless they are in a part of their range that is so arid that there is nothing to browse. They are inefficient eating machines. The critical concern with our elephants, though, was that they had also stopped drinking. The enemy was not starvation. It was dehydration.

We are not certain exactly how much water an adult elephant takes in each day, but we estimate that it is about twenty to thirty gallons. Despite the fact that their water intake was down to practically nothing and they were not eating at all, the three elephants with the worst symptoms were defecating almost constantly. This meant that the walls of the barn, up to a height of about ten feet, and the entire floor, had to be washed constantly.

The barn, then about five years old, has since been replaced by a completely new elephant holding area, designed to support a much larger herd. Back then the barn had two relatively small stalls and one much larger stall. Giant steel columns allowed keepers to pass easily into the elephant area but precluded the animals from walking out. The floors were cement and the walls cinderblock, but the floors themselves were heated for comfort during the winter. There were large skylights and two off-exhibit yards where the elephants could go to exercise. Their exhibit was very large, but the staff did not like to work with the elephants while they were on exhibit. The theory was that the exhibit yard should remain a place where elephants behaved in a natural fashion, free from the demands of the training staff. Water was provided at four spots along the edge of the stalls and was available on demand twenty-four hours a day. When the elephants became sick, cleaning the walls was problematic, partly because the fecal material was often too high to reach and partly because no one wanted to disturb the sick animals.

At first they just had loose stools, but by day three they had what could only be described as projectile diarrhea. Black-green and almost pure liquid, their diarrhea was nauseating. The smell was overpowering, and David Hagan, the head keeper, remembers that he had to close the barn's huge metal doors against the cool night air, which trapped the overwhelming odor inside. As soon as he finished cleaning one giant mess, the next elephant would blast diarrhea all over the walls and floor. All night long he moved quietly along the line of elephants with a squeegee, pushing the mucous-laden, bloody liquid toward the floor drain. The biggest problem, though, was not cleaning up the fecal material (which was absolutely essential if we wanted to avoid reinfection). The biggest problem was giving them fluids.

As soon as we made an initial diagnosis, the staff was on the phone lining up bags of glucose/saline solution, which needed to be given intravenously. Elephants use their ears like giant radiators to cool their bodies. If you look at the back of an elephant's ear, you'll see a pattern of very large veins. The blood circulates through the ears and is cooled before returning to the heart. It is relatively easy to give an African elephant an IV because the veins are so prominent and the skin behind the ear is surprisingly soft and supple. The hard part is keeping the intravenous needle in. Human patients, in a hospital, aren't prone to flap their arms and dislodge an IV, but elephants flap their ears regularly. Installing an IV meant putting a catheter in the vein, gluing it down, stitching it to the ear, and then taping it down to protect the stitches and hold it in place. Gauze, covered with tape, held the whole thing down. At first, the elephants tried to feel for the catheter with their trunks, but eventually they became so lethargic that they stopped that as well. Still, the vet staff were kept busy reinstalling catheters on an average of once every eight hours or so.

On top of that, an elephant's ears are about eight feet off the ground. To administer an IV, we had to have a keeper walking alongside the elephant, holding a ten-foot pole over the elephant's head so that gravity would feed the solution into her body. This needed to be done twenty-four hours a day for about five days. It was exhausting work and could only be done by one of our regular staff. At the height of their illness, the lethargic elephants were receiving in excess of one hundred liters of fluid every twenty-four hours. The keepers eventually became so exhausted by the round-the-clock effort that they sat slumped in chairs while they held up the IV poles. Debbie remembers falling asleep in a chair next to Kubwa and waking up sometime later, still holding her pole above Kubwa's head.

In addition to rehydrating the elephants, we were treating them with antibiotics. Twice a day the vets would take a six-inch syringe, an inch and a half across, and give each elephant an injection. They rotated from leg to leg. The two drugs used, first Amikacin and then Ceftiofur, were very powerful and had never been used on elephants before. As a consequence, the vet staff also had to draw blood in an effort to monitor for possible kidney or liver damage. The antibiotic course ran for fourteen days.

Having the staff involved in the treatment meant that we had to have volunteers to do the cleaning and, as I said, there was a lot of cleaning to be done.

My wife, Melody Noel (we had long discussions about changing last names upon getting married but, in the end, I stood fast and refused to change mine), became a volunteer. Like many people who spent any time at all around these animals, she had fallen in love with them and was deeply concerned about their health.

I was also concerned, in part because I identified closely with what I imagined them to be going through. While in graduate school, I had spent two months in Tunisia, traveling to different types of oases—desert, forest, and coastal—to study how agriculture varied from place to place. I thought that I might find something that would make a better dissertation topic than working on land reform in India. I didn't, but I did have a fascinating and wonderful trip. Wonderful, that is, until the very end, when I contracted dysentery. I came down with the initial symptoms on the boat that brought me back from Tunis to Marseille. By the time I reached Paris, I was planning my every move according to how far I would be from the nearest toilet. I had tickets for the hovercraft, which meant a bus ride from Paris to Calais (a ride of perhaps 150 miles), a quick trip across the English Channel, and then another bus ride from Dover to London after clearing customs.

By the time I reached Dover, I was actually concerned about getting through customs. I had heard from someone that the English were funny about letting sick people into their country. After all, they had socialized medicine, which meant that they had to treat everyone for free, including tourists. It would obviously be cheaper and easier just to not let sick people in to begin with. And I was sick. My eyes were sunken, my face was pale, I was feverish, and I know that I kept eyeing the customs line and plotting a quick dash for the closest loo. If I didn't look sick, I must have at least seemed suspicious, but they let me through without a question and I boarded the bus to Dover, only to discover that the English bus, unlike its French counterpart, did not come equipped with a toilet.

I think I would have made it to London, and then to the airport and home, if the bus had a toilet, but it didn't. Maybe I would have made it anyway, even with no toilet, but we'll never know. As it happened, not far out of Dover, the bus broke down. At that point, desperate, I asked the driver if I could slide down the roadside embankment, as I "wasn't feeling well." I took some Kleenex and, squatting behind some bushes, made a horrible discovery. I was no longer defecating any fecal material at all. Everything was blood.

Anthropologists will tell you that, of all the fluids that occasionally leak from humans, including urine, spittle, vomit, mucous, and tears, only two fluids are widely thought to contain our life force—semen and blood. Beliefs regarding semen as a life-force fluid are not really universal, although, where you find them, they are fascinating. For example, in his book *Guardians of the Flute,* Gilbert Herdt describes the Sambia, a group of people who live in Papua, New Guinea, that practices ritualized homosexuality. They believe that the males from the older generation must endow the next generation with life-giving sperm, even though they themselves lose part of their life force in the process. Ideas about blood as a life force are certainly found in our culture and find expression in many ways. We're related "by blood," become "blood brothers" by commingling our blood, and express our depth of feeling where blood is concerned in any number of other ways.

I doubt I was thinking about any of that at the time, but I do know that, at that moment, I lost the will to go on. I walked back up the slope and told the driver that I was much sicker than I thought. Fortunately, he had a radio, and within fifteen minutes an ambulance pulled up behind the bus, and I was on my way to a hospital. I put my gear in the back and rode, sitting up, all the way. At the hospital, though, my emotional state took another severe blow. The admitting staff asked only a few questions, and their exam was little more than cursory. They asked where I had been, what I had been doing, took my temperature and blood pressure, and announced that I would have to get back in the ambulance for a trip to a specialty hospital for infectious diseases.

This time the ambulance drivers weren't as talkative. They had also put on masks; the initial diagnosis was typhoid fever.

I've since discovered that medical personnel will assume the worst when otherwise not uncommon symptoms manifest in someone who has been in an unusual place. For example, shortly after my return from India, I caught a bad cold and tore my diaphragm during a particularly violent coughing fit. The young doctor at the teaching hospital at the University of Michigan was thrilled to announce that I might have liver flukes. Fortunately, calmer heads prevailed.

But I hadn't made that discovery yet, and I was very worried. I was also getting sicker and sicker by the minute. I lay down and tried to sleep. It didn't take long after my arrival for the good people at the next hospital to conclude that I didn't have typhoid fever. Interestingly, though, they had already burned all of my clothes when they finally decided that I wasn't carrying the next plague. I may not have had typhoid (instead it was just a very bad case of dysentery), but I was definitely very sick. Sick enough, in fact, not to care, when shortly after getting to my room (a beautiful private room in the isolation ward), a nurse very gently asked me what religion I was and whether or not I had any messages that I might want to leave for my family. I wound up staying in the hospital for almost four weeks, all courtesy of the British government. For the first three weeks, I didn't even try to get out of bed.

Anyway, what I remember the most about all of this is the feeling I had when I gave up—empty and hollow. Sometimes people really do just let go of the will to live. It is not something I would ever want to do again.

To this day, I think that after a certain point our elephants just gave up. When they first got sick and stopped eating and drinking, David Hagan recalls running to the grocery store and filling carts with anything he thought might tempt them. He bought every manner of fruits and vegetables, pastries, donuts, and even Pop-Tarts, Gatorade, fruit punch, and Kool-Aid. As the mini-epizootic reached full proportion, the focus shifted to maintaining the IVs around the clock. It was not until ten days after the symptoms were first noticed that the staff again tried to get them to eat and drink. The illness had run its course, but it had left three very disheartened elephants in its wake. Those three, Cita, Kubwa, and Ivory, simply were not interested in living.

Ivory looked the most wasted. The smallest and youngest to begin with, the weight loss had affected her dramatically. Each side of her head showed a great indentation, and the flesh over her hindquarters had withered to almost nothing. Debbie was on the phone to elephant handlers around the country, concerned that Ivory wanted to lie down and worried that her girl would not have the strength to get up again. The lungs of an elephant do not float in the rib cage in the same way that they do in humans, and it is very hard for an elephant to breathe while lying on its side. If an elephant were to lie down and not have the strength to get back up again, its pulmonary functioning would be severely compromised and death would likely follow. Eventually, Debbie decided to build a gently inclined ramp of sand out in the exercise yard, reasoning that if Ivory wanted to lie down it would be easier for her to get back up. Ivory eventually did lie down for short periods—perhaps the rest helped.

In less than two weeks, the illness had run its course. The antibiotics had kicked in, the diarrhea had stopped, and the elephants had taken enough liquid through their IVs to stay alive. But Ivory, Kubwa, and Cita were still not eating or drinking. A few days before, staff had again begun the process of tempting them with everything they could lay their hands on, but nothing had worked. The weather had improved, it was sunny and warm, and the animals seemed to enjoy standing in the warmth outdoors, often leaning against the side of the barn to rest. But they would not eat. Our vet suggested that, if we could get them to eat anything, IronKids bread would be the best. They had to be deficient in iron by then, and if the commercials were to be believed, IronKids had to be the best thing for them.

My wife was still volunteering. There wasn't much to do, though. The barn had been thoroughly disinfected, and the exercise yard, exposed to the bright sunlight, tested negative for the bacteria. Most of the time she joined the staff out in the yard, and they were all increasingly disheartened at the elephants' lack of interest in eating and drinking. "It was like they never wanted to see food again," she said, "almost as if they were nauseated by the thought." Inside the kitchen, adjacent to the exercise yard, fifty loaves of IronKids bread stood as a mute reminder of their concern.

At that point, they had been trying for three days to get the elephants back on feed, but the fourth day would prove the charm. Debbie and my wife were sitting on a large tire (used as an elephant toy), inside the exercise yard by the kitchen door. They had been trying, with no success, to feed Kubwa a piece of bread. They would hold it out, Kubwa would politely sniff it with her trunk, then let it drop to the ground and slowly move away.

Now an elephant's trunk is an amazing organ. It is strong enough to push down a tree yet delicate enough to pick up a dime. It is also incredibly juicy, as noses go. Eventually Debbie and my wife had accumulated a wet little pile of bread. They began to idly form the wet bread into round balls, tossing them up and down in their hands. They did not notice as Kubwa moved toward them, but in a magic moment a trunk reached out and caught one of the balls in midair. Kubwa popped the ball in her mouth. A look of astonishment crossed their faces, and soon both women were busy making bread balls and tossing them to Kubwa, who began to eat like she hadn't eaten in days. In fact, she hadn't eaten in days. Attracted by the activity, Ivory and Cita wandered over. Cita was not one to let another elephant have something that she couldn't have, and Ivory, never one to be left out, soon wanted her share, too. For the next hour the women made bread balls and tossed them to three suddenly hungry animals. From that point on, they gained weight and strength every day. Whatever malaise had affected them was gone.

We never did tell the good people at the Sara Lee Corporation that it was IronKids that broke the long and anxious fast.

3

The Tragedy in Philadelphia and the Miracle in Chicago

Pete Hoskins, the director of the Philadelphia Zoo, is in bed, sound asleep, when his phone rings. It is past midnight, so technically, it is the earliest hours of Christmas Eve, the morning of December 24, 1995. On the line is one of his security personnel. "There's been a fire in the World of Primates," he is told. "You've got to get over here." Whatever he has been dreaming, it is nothing like the nightmare he will find now that he is awake.

He rushes to the zoo and then to the primate house. His curator of primates, Dr. Andy Baker, is already there, having beaten him by a matter of minutes to the scene of the fire. Karl Kranz, the general curator, was also on the scene, but Pete was focused on Andy. He remembers looking into Andy's eyes, full of pain but not tears, and hearing him say, "They're gone. They're all gone." All of the animals in the building—the gorillas, the lemurs, the orangutans, and the gibbons—all twenty-three of them are dead.

Now, eight years later, Pete and I are sitting in Hank's Restaurant in Charleston, South Carolina, after a long day of sessions at our annual zoo and aquarium directors' retreat. I've heard most of this story before, but I've prevailed on Pete to tell me the whole thing from the very beginning, not just about the fire but what happened to the zoo before and afterwards. It's been a long time since Christmas Eve of '95, but I figure this can't be the easiest conversation for Pete to have, so I suggest we split a bottle of wine over dinner. As it happens, I'll never see a drop of that wine. In fact, when he's done telling me this story, I'll see him get up slowly from the dinner table, go to the bar, and order another drink. And I won't blame him a bit.

To this day, Pete Hoskins looks young. He has that kind of friendly, boyish face that never seems to age, and he always seems to be smiling. He graduated with a master's in public administration and moved to Philadelphia in 1972 to work for the city. He didn't really become interested in zoos until he became director of Fairmount Park, a job he held from 1980 until 1988. The zoo opened in Fairmount Park quite a bit before Pete's arrival on the scene. In fact, the Philadelphia Zoo opened to the public on July 1, 1874. It is America's oldest zoo.

As head of the park, Pete worked closely with Bill Donaldson, the zoo's dynamic and innovative director. They stayed close even after Pete became the city streets commissioner, a job he held until 1993. Bill Donaldson, Pete's close friend, died in November 1991, leaving the zoo without a director. A member of the zoo's board, Marsha Pearlman, graciously stepped in as interim director. By January 1992, Philadelphia had a new mayor, and a national search for a permanent new zoo director was under way. Pete was encouraged by board members to apply, and he did—but he didn't really think he had a chance for the job. Still, he did his homework. He studied the zoo's master plan, visited zoos around the country, and read everything he could get his hands on.

The most important thing he did was to talk to other zoo directors, and the most important zoo director he talked to was Bill Conway, the legendary head of the Bronx Zoo or, more precisely, the Wildlife Conservation Society (WCS). Bill had started his career at the Saint Louis Zoo as a bird keeper, but he rose rapidly through the ranks. By the time he landed in New York, he was not only one of the most knowledgeable and progressive of the American zoo directors, but also one of the most charismatic fund-raisers in the nation. In the years he headed the WCS, he grew their conservation endowment fund from next-to-nothing to some four hundred million dollars and made WCS the single most important force in the world for field conservation work. Pete actually hadn't gone to see Bill. He figured it would be nearly impossible to get an appointment. But somehow Bill learned that Pete was on the grounds of the Bronx Zoo, dropped everything he was doing, and spent the rest of the afternoon driving Pete around the zoo, sharing his vision of what zoos are and, more importantly, what they should become. Pete says that, when it comes to zoos, it was the single most important conversation he's ever had.

By the fall of 1993, almost a year after first applying for the job, Pete suddenly found himself the new director of the Philadelphia Zoo. At his final interview, what probably got him the job was, first, that he had a vision for what the zoo could be (thanks, in large part, to Bill Conway) and, second, he felt that the existing master plan, which called for a five-million-dollar expansion, was way too timid. As soon as he got the job, he doubled the capital campaign goal.

By mid-1994 he had raised nine million dollars, but he recalls that not everybody seemed very happy about it. In fact, some of the board members seemed downright resistant to his efforts. But Pete continued to think even bigger. By the fall of 1995, he had applied for a major planning grant that would result in a far larger campaign than the one he had already doubled in his first year on the job. By mid-December, just days before the fire, the zoo was awarded a Master Plan grant from the William Penn Foundation, positioning Pete and the zoo to catapult ahead. He announced his plans at the December meeting of the board and told them he'd come back by the January meeting, ready to start the planning process.

But not everything was going smoothly for Pete. In the month before the fire, he determined that he had to fire a very popular curator at the zoo, who, he felt, was just not buying into his new ideas. Shortly after that, Bill Konstant, another key zoo employee, resigned. Both events were reported in less than flattering terms by the press, and the board, or at least some of them, seemed resentful. Still, as Christmas approached, Pete was looking forward to an exciting and eventful new year, thanks to the William Penn grant.

Then came the phone call in the dark hours of the early morning. The first thing he saw when he got to the zoo was all of the fire trucks and hoses. The building was made of cinder block and it appeared to be perfectly intact, but the smell of smoke was overwhelming. The red lights of the fire trucks cast an eerie, spinning glow as Pete made his way to the holding area in the back of the building. It was there that he met Andy, there that he learned that his animals were dead.

Pete is one of those people who can smile even when they're crying, and he smiles at me now. "I kicked a fire hose," he says. "You have no idea how hard a fire hose is when it's under pressure. It was like kicking a rock." He knew he had to see for himself what had happened, so, with the aid of a flashlight, he moved slowly down the corridor to the first holding area. The first animals he saw were a mother orangutan, Rita, with her baby, Jinga Gula, curled in her arms. "They looked so comfortable," he told me, "so peaceful—

almost as though they had died in their sleep, not in struggle but in repose. They were covered with soot and ash, sort of a dull gray by the light of my flashlight, but otherwise they looked just as they would have if I had tiptoed in to look at them in the middle of the night." Later, Pete's initial impressions would be confirmed. As they carried each dead animal out of the building, they would find, left behind, the perfect outlines of their bodies on the ash-covered floor, even down to the light fringes of delicate fur. There had been no struggle, no fierce fight for life. They had not been killed by fire. They had, ever so slowly, been suffocated by smoke.

Pete now tells me that he was relieved to have been one of the first to see what had happened. He recalls thinking that somehow it would help prepare him to deal with whatever lay ahead. Oddly, he can't remember what he did next. Did he hug Andy? All he remembers is thinking that the world had changed for him and that whatever he did next would have a dramatic effect on how his staff, his supporters, and the people of Philadelphia would deal with what would soon come to be regarded as one of the worst fires in his city's history and, undoubtedly, the worst tragedy involving animal deaths in the history of American zoos.

Pete, Andy, and Karl moved to the zoo's Animal Department building and were quickly joined by the zoo's director of public relations, Antoinette Maciolek, and later by Fire Commissioner Harold Hairston. Meanwhile, the zoo's phone tree had been activated but, of course, not everyone could be reached, due to the holiday. Clark DeLeon, in his wonderful history of the Philadelphia Zoo, *America's First Zoostory,* describes how the word went out:

> Some Zoo people were surrounded by loved ones and visiting relatives sleeping in spare beds during the holiday when they were awakened by a telephone call in the middle of the night. Others heard the news on the car radio on the way to work the next morning. Others arrived at work and wondered why there were so many TV news vans around the Zoo so early on a day when the Zoo was closed to the public. Others heard the news on TV as the family sat down to Sunday breakfast, their children bursting into tears over their cornflakes. Others were on vacation elsewhere when CNN or a grief-stricken relative calling on the phone interrupted their holiday with the terrible news. And others were there before the firemen had rolled up their hoses.

Pete realized that the information that reached the world through the Philadelphia media gathered outside the building had to be as accurate as possible, but there was so much the four of them couldn't possibly know. By 2:00 AM they were joined by the fire commissioner, who told them that, in his experience with fire deaths in homes at night, the cause of death is almost always smoke, not flames. He told them that the fire marshal had taken charge of the scene and that, although the investigation would take time, it appeared as though the fire started in the ceiling and was caused by an electrical short. Ironically, the primate house was the second newest building in the zoo—one with modern and up-to-date wiring.

Karl called the Bronx Zoo, realizing that all of the animals would have to be removed from the building and necropsied (a necropsy is an animal autopsy). Most people probably don't know this, but every animal that dies in a zoo, even if it's not part of the collection and flies, hops, or crawls in on its own, gets necropsied. In 1995 only five zoos, however, had full-time pathologists on staff—the Bronx, St. Louis, the National Zoo in Washington, and San Diego. The fifth zoo was the Philadelphia Zoo. In fact, Philadelphia was the first zoo in the world to have a full-time pathologist, and much of what we know about the causes of mortality in animals has come from the Philadelphia Zoo. But the zoo's pathologist, Dr. Virginia Pierce, would not be able to necropsy so many animals at one time, and the Bronx Zoo was the closest big zoo that could spare their pathologists to help with the huge task ahead.

Next they arranged to have the animals moved to the stainless steel walk-in cooler adjacent to the necropsy lab. To this day, Pete doesn't know all who participated in this awful task; he knows that Ken Rebechi, on the maintenance staff, was one, and that his chief financial officer, Matt Schwendermann, was another. You have to understand that gorillas and orangutans are very heavy. Male gorillas can weigh between 350 and 500 pounds, and male orangutans can weigh upwards of 300 pounds (with the females of both species weighing about half as much). Moving them is incredibly hard work, and conducting a thorough necropsy on an animal so large could take five or six hours. With twenty-three primates, the task was even more daunting.

But before that process could begin, something even more important had to happen. The keepers who were closest to the animals had to have a chance to say good-bye. Perhaps no one was closer to them than Julie Unger-Smith, the lead keeper for primates. I'll let Clark DeLeon tell her story:

> She had been spending the night at her fiancée's family's house when the call came. "I was pulled out of bed and brought to the phone," she says. It was Andy Baker. "I heard later he was absolutely tormented about how

to tell the keepers," she recalls. "He's always been a very straightforward person and there was no easy way to say it. He just said, "Julie, something bad has happened . . .'"

Tears overtake her story. She recovers and continues. "Which one? Which animal?' I said to Baker. I knew he couldn't be calling to say that one of the moats was overflowing or a rail broke. I knew someone had died. I absolutely knew. And I knew it was an animal. Looking back, if it had been one of my co-workers, it would have been devastating. But I knew it was an animal. But which one? Oh, God, which one?

I just couldn't comprehend what he said. He said, "There was a fire and . . . all the animals died at the World of Primates." And it's so goofy now, but I remember saying, "All of the gorillas?" And he said, "yes." And I went through all the species. "All the orangutans?" And he said "yes." "All the gibbons?" And he said, "yes." All the lemurs . . .

At the same time that Andy was calling the keepers, Pete was calling other key staff and the chairman of the board, Bob Wolcott. Bob agreed to start calling other board members, and more importantly, he rushed to the zoo so he could be there at daybreak for the first press conference.

By 6:30 in the morning, Pete, supported by Bob and Antoinette, was ready to talk to the media. They still didn't have the answers to many of the most obvious questions they would be asked. They weren't sure how the fire started and couldn't say if or when the building could be reoccupied. In fact, they weren't sure they would want to even if they could. They knew the protocols, but there was no way they could anticipate the answers to every question. Pete remembers being asked, for example, what would happen to the animals after the necropsies. As it happened, Karl had already made arrangements for their remains to be cremated, but Pete was left speechless by the insensitivity of the question. They had already decided that the zoo would remain closed for the entire week to come to allow everyone an opportunity to grieve.

Of course, the zoo was flooded with visitors anyway. Staff members who would normally be on vacation, volunteers, donors, all came in the next few days. Gradually it dawned on Pete just how many intimate connections so many people had to so many animals. Keepers who had seen some of those animals born, volunteers who had come to daily observations, vets who had come in for special procedures. "How can you understand," he asked me, "the hundreds, maybe thousands of special connections that so many people had with so many animals?"

Pete had his own special connections. Shortly after he became the director, he had been invited down to watch a veterinary procedure on one of the gorillas, Kola, who needed her knee X-rayed. He recalls looking into Kola's eyes and thinking, "She looks just like a frightened little girl would, if she were being sent to some strange and scary place like a hospital, not fully understanding what was going to happen to her or why." Pete looked into the distance and said, "That was the moment I really felt that I was as responsible for that animal as I was for my own family." Now, his voice barely a whisper, he says, "Kola died in the fire."

And then there were the donations. By Christmas Day they began to flood in, unsolicited. Pete remembers opening a letter from a young girl named Joey. Her scrawl read simply, "I'm sending you all my money." There were fourteen one-dollar bills carefully folded inside. In the months to come, the zoo would receive over fourteen thousand gifts, some of them in the form of bags of pennies.

In the hours after the fire, the staff decided that they would open the zoo for free for the entire month of January, but they still hadn't figured out what, if any, memorial service they should have. At 5:00 AM, two days after the fire, Pete called a friend, Sandy Bressler, from the Arts Commission, with the idea of creating a memorial gallery wall. "Can you do something beautiful in the next three or four days?" he asked her. "Something that captures their spirit and their lives? Something that tells the stories of the people who cared for them?" She said she would do anything to make that happen. Miraculously, the art community of Philadelphia, working with zoo staff and volunteers, created just such a memorial. It was completed by the day after New Year's, when the zoo once again reopened to the public.

If he hadn't figured it out by the end of that long Christmas Eve day, it would become clear to him the next day that everything had changed. By late morning on Christmas Day, his family still hadn't had an opportunity to open their presents. By noon, the family sat down together, only to have the phone ring again. Yet another news station was calling and this one wanted to do an interview in his home. His oldest daughter put a hand on his shoulder and said, "Dad, it's okay. I understand."

"Funny," says Pete, "where your most important support comes from."

But in the days and weeks following the fire, Pete would discover that he most unequivocally did *not* have everyone's support. The zoo was investigated by the city controller, the zoo board called for an investigation, and the American Zoo and Aquarium Association (AZA) sent an accreditation team.

The *Philadelphia Daily News* wondered if he should be allowed to keep his job. Even Pete began to doubt himself. In Julie Unger-Smith's words, "Pete Hoskins actually said to us later, 'We let the animals down.' Not that we were taking the blame for doing any one thing wrong, but that they were in our care when this happened. Pete genuinely felt the pain, for himself and for us." The fire, in many ways, changed everything. Says Clark DeLeon, "The Zoo was tested as never before, and no one was tested more personally than Pete Hoskins. The second guessing, whisper campaigns, and outright accusations of lax management began soon after the fire, and none so forcefully as in the press. Hoskins had enjoyed a relatively favorable relationship with the city's news media since entering public life, but now he found himself on the receiving end of harsh criticism."

In the three months that followed, Pete pushed hard to have a new master plan created for the zoo, but the internal strife on the board precluded any kind of consensus. Finally, Pete was called to a half-day board retreat. His board chair let him know that his job hung in the balance. The biggest issue, related to the planning process, turned on the question of his vision of the zoo, now that the fire had happened; what did he propose to do with the primates; what was his master plan? He knew that he wanted to break new ground—define what the next step in animal care and exhibition would be—not just for primates but for the whole zoo. And there was just no feeling of unanimity behind him.

It would have been interesting to be the proverbial fly on the wall during that meeting. I, of course, wasn't there, so we'll have to take Pete's word for it when he says, "It was the best performance of my life." The board eventually endorsed Pete and his new plan, and the harshest critics on his board quietly resigned over the next few weeks. The zoo would rebuild and the hearts of its supporters would mend, but everyone knew that nothing would ever be the same again for the zoo. As DeLeon says, "In all of the best of times and worst of times that the Philadelphia Zoo has enjoyed or suffered, none was quite so exquisitely painful and enormously inspiring as the fire and its aftermath."

On August 16, 1996, almost nine months after the fire in Philadelphia, a miraculous event occurred at the Brookfield Zoo in Chicago. A three-year-old child fell almost twenty-five feet down into the zoo's gorilla exhibit. What happened next was both terrifying and, in a gentler way, almost as inspiring as the response to the fire in Philadelphia.

The story plays out in a span of nineteen minutes, beginning at 2:10 PM on a busy Friday afternoon at the zoo. This time it was not a phone call but a burst of radio static followed by a terse message that set events in motion. The message was, "Base from 409, we have a signal thirteen in Tropic World—Africa" ("signal thirteen" was the zoo's code for a life-threatening situation). "There is a child in the gorilla exhibit. I repeat. There is a child in the gorilla exhibit." Just over nineteen minutes after that message was sent, the child, conscious but in critical condition, would arrive at Loyola Hospital. He would recover fully, but according to his mother, would have no memory of the frantic efforts of the staff to save his life. Nor would he remember the kindness shown to him by a western lowland gorilla named Binti Jua, whose actions may have played a far more important role in his rescue than any one thing that the staff could have done.

Binti Jua was born in the San Francisco Zoo in 1988 and came to the Brookfield Zoo in February 1991. Her name means "daughter of sunshine," and although she had a wonderful temperament, she was not an experienced mother. Binti Jua had been hand-reared in San Francisco. The staff at Brookfield brought her to their zoo with the hope that, by integrating her into a stable female population, she might learn from Brookfield's breeding females. The Brookfield females were all related, but they had lost their magnificent silverback, named Samson, to a brain tumor in 1988. After trying to bring in other males to replace Samson, the zoo decided to take a chance and bring in a very old male that had never reproduced before and, in July 1992, a thirty-six-year-old male named Abe was brought over from Omaha's Henry Doorly Zoo.

While Abe was definitely a silverback and the Brookfield females felt his presence instantly, Abe really bonded with Binti Jua. In fact, Binti was the only female he ever bred with.

When the staff at Brookfield learned that Binti Jua was pregnant (at the relatively young age of six), they immediately set out to teach her parenting skills, using dolls to help her learn how to hold and nurse an infant. Not only was Binti Jua a very young first-time mother without much interaction with other mothers and their infants, but she also lacked any personal experience with mothering since she had been reared by humans. The staff hoped that, when the time came, Binti Jua would exhibit her basic maternal instincts (helped, of course, by a little coaching and training from the staff).

In 1995 Binti Jua successfully gave birth and began raising her daughter, who was named Koola. The combination of instinct and the keepers' prenatal training had paid off. (By the way, Koola thrived and is now, at the age of ten, caring for and raising a beautiful daughter of her own named Kamba).

Binti Jua was always a relatively low-ranking female in Brookfield's group. This was, in part, because she is not very big by gorilla standards, weighing only about 165 pounds, but mostly it was because the other four females in the group all had the same mother, named Alpha. Those four related females formed a fairly tight bond, and Binti Jua was always somewhat of an outsider in the group, even after the birth of her own baby.

Brookfield's gorillas live in a giant building, "Tropic World," which features primates from South America, Asia, and Africa. Visitors enter the building at a fairly high level above the exhibits, but since many primates live in trees, they find themselves eye-to-eye with a wide variety of species as they move through the three separate areas of the skylighted building. At the far end of the exhibit is the area devoted to Africa, and here the trail winds around a large gorilla exhibit dominated by a twenty-foot-tall rock mountain, located in the center of the exhibit and topped by five huge artificial trees. Visitors can circle all the way around the exhibit, and the pathway is much lower in the front of the exhibit than it is in the back.

Around the perimeter of the exhibit there is a continuous series of planters, filled with vegetation that grows over the edge of the steep walls around the gorilla enclosure. The whole thing is really a giant pit, with a mountain in the middle. When the gorillas are on top of the mountain, resting or climbing in the trees, they are actually above the eye level of visitors, but it is a long drop to the open area around the base of the mountain from any point around the perimeter.

There is also a bamboo fence that is cleverly designed to keep visitors from climbing into the planters (and from potentially falling down into the exhibit), using short tree branches roped onto metal railings designed to resemble bamboo. In most areas, the tree branches make it virtually impossible for an adult to climb the rails. The whole exhibit is designed to have steep enough walls to keep the gorillas in, yet still provide great views for visitors. A small stream flows partway around the base of the mountain, and there are a variety of places for animals to rest, climb, socialize, or be alone. On the whole, it is probably the single most impressive exhibit of the three major sections in Tropic World. It is almost certainly the most popular.

The little boy, whose name was never released to the press (and can't be given here, either), was visiting the zoo with a group of about eight other children and several adults, including his mother. He had run ahead of the group with his older brother and sister, excited to see the gorillas at the top end of the long path through the building. As is always the case on a nice summer day, the zoo was crowded, and it is easy to see how, first, a mother could lose sight of her child for an instant, and second, why she wouldn't be too worried about it. After all, the pathway ends in a giant circle around the gorilla exhibit and the children would probably have had to meet their group coming back, even if they got very far ahead of the group.

I'm not even sure if the boy's mother ever completely lost sight of him. What I do know is that, when he ran forward to join his brother and sister (who were leaning up against the railing of the exhibit), his momentum carried him right over the top of the railing and down into a planter just below—a planter located on the farthest (and highest) side of the exhibit. Visitors across the exhibit could see the prone child teetering on the edge of the planter. At least one of the visitors had a video camera and taped most of what happened, but many other visitors had cameras and photographed much of what happened next. There was a collective gasp as someone pointed toward the child and screamed. He seemed to hang on the edge of the planter for an instant or two, and then he fell, bouncing once against the steep wall of the exhibit and landing, thank God, more or less on his butt.

Hundreds of visitors saw the whole thing. Many ran out of the building, some panic-stricken and some looking for someone to help, and that was when the staff of the zoo jumped into action. Normally, there are at least ten or eleven of the fifteen full-time keepers in Tropic World working in the building at any given point during the day. On that afternoon, seven of the keepers had taken their lunch later than normal because of the unusually large crowds. These keepers were together at a picnic bench not far from the doors to the Africa exhibit. They were walking in a group back toward the service entrance to Tropic World when the crowd began to pour out of the building's emergency exit. This meant that were seven keepers who were perfectly positioned to respond to the crisis. Of them, lead keeper Craig Demitros was the first to look into the gorilla exhibit. It was Craig who made that initial radio call.

Craig immediately took charge of the situation and sent three keepers down to the lower level of the building. At the lower, non-public level there are two doors that connect the daytime exhibit space with the overnight, or off-exhibit, holding area. The doors are operated electronically, and Craig needed those three keepers to open the doors so that gorillas could be taken into the holding space. Meanwhile, after charging through the excited crowd and radioing the all-staff alarm, Craig looked down from the public walkway just in time to see Binti Jua move her baby, Koola, onto her back and gently pick up the unconscious child.

Craig and the other two keepers stayed on the upper level and, following the emergency protocols developed for just such a situation, grabbed the water hoses used for cleaning the exhibit. The hoses were located at three roughly equidistant points around the exhibit. This was, I guess, their second lucky break. They had, just a week before, walked through a "visitor falls into the gorilla exhibit" drill. They knew precisely what to do and where to go. More importantly, because they had practiced, they were able to coordinate their efforts without any ability to communicate or signal to each other. There was simply too much noise from the panic-stricken crowd to shout over. Now you could say that there was nothing lucky about it. In fact, it really was plain, old-fashioned good planning and cool-headed professionalism on the staff's part. But, on the other hand, you have to admit that it was at least a happy coincidence that they had held the drill just days before.

The three keepers shot their hoses down at the floor of the gorilla exhibit, a signal to the gorillas that they should go into their overnight, or off-exhibit, holding area. Binti Jua went toward one of the doors, and the rest of the group went to the other door. This was normal for the Brookfield gorillas. Binti Jua always used a separate door from the rest of the females. Now comes the staff's third lucky break. Binti Jua, knowing she should go in, gently sets the boy down outside the door, while keeping her own baby on her back. She then walks around to the other side of the mountain, where the other gorillas are, and joins them as they go into *their* door to the holding area. Here's their fourth lucky break. That day there was no silverback on display. Had there been, everything might have turned out differently. A silverback *might* have been more inclined to be aggressive toward what might be perceived as a threat (a person jumping or falling into "their" territory). But there is little question that the real lucky break was not so much that the silverback wasn't on display so much as it was that the child was unconscious. The staff at Brookfield feels that the worst-case scenario is if someone falls into the exhibit and remains conscious. Their protocol in such a situation is to try to calm the visitor and direct him or her to sit or lie quietly, to not move, and to keep face and eyes averted from the animals. This sounds good in theory, but it would be hard enough to do with a panicked adult and probably almost impossible to do with a young child. With an unconscious adult or child, however, the gorillas would not perceive that there is anywhere near as much of a threat posed, and hence would be much less likely to display defensive aggression.

Within a few minutes the doors have opened, the gorillas are off the exhibit, and the boy, now beginning to stir a little, is safely on the ground, accessible to the staff. Now comes the fifth piece of good luck. As it happens, many of Chicago's paramedics are in the zoo that day for a group picnic. The Brook-field Zoo has its own ambulance, and its security staff has EMT training, but paramedics have more training as first responders. Two of them, having seen almost the entire series of events, follow down to the lower level to assist. Their presence was, perhaps, a little less outright good luck and a little more of a mixed blessing. They didn't follow the keepers' instructions to hang back and, by bursting into the holding area, they frightened the gorillas. The staff, though, was able to deal with this momentary problem.

I know this is starting to sound like winning the lottery, but the sixth piece of good luck was that the senior staff, including the vet and the curator of primates, Melinda Pruett-Jones, were having a meeting in the Animal Hospital's conference room barely one hundred yards from Tropic World and, as always, they had left their radios on. As soon as they heard the code thirteen call, they (in Melinda's words) "levitated," leaving the zoo's rather startled CFO wondering what she had said to cause everyone to run out on her. The zoo's chief vet, Dr. Tom Meehan, and one of his veterinary technicians jumped into the vet truck, and while Tom drove through a crowd of visitors walking between exhibit buildings, his technician calmly loaded tranquilizer guns in case they needed to immobilize an animal. They arrived just as the gorillas were locked down, and so, while the guns weren't needed, they were at least available to provide medical supervision, along with the EMTs and para-medics.

Melinda brought the child's mother down to the lower level. Tom had al-ready gone down below to the exhibit access door. The visiting paramedics, the staff EMTs, and the keepers went into the exhibit and secured the child to a board, as they would with anyone who had potentially injured his spine. Tom was able to tell the mother, who just happened to be a nurse, that her child was crying and talking a little—both very good signs. Then the mother joined the paramedics and the EMTs driving the zoo's ambulance for the short ride to Loyola Hospital. When the ambulance arrived they clocked in at just over nineteen minutes after the first radio call went out.

Tom did not go along with the ambulance, which was actually pretty full by that point. Instead, he went to where the zoo's security staff was inter-viewing witnesses in order to write an incident report. Tom had heard they found the visitor who had taped the whole thing, and Tom very much wanted a copy of that tape. The visitor gave a security officer the tape to copy and he passed it along to Howie Greenblatt, the zoo's videographer, promising to return the tape in a few minutes. The tape was later released to the media, but not by the zoo. It was released by the man who actually shot it, and as we'll see in a minute, it was aired all over the world.

By the time Howie was done copying the tape and had returned it to the owner, Tom and Melinda were meeting with the zoo's public relations staff to prepare for the media onslaught. They knew it was coming, if only because there were news helicopters circling the zoo even before the little boy left in the ambulance. Howie, holding a copy of the tape, burst into the meeting. Tom remembers him saying, "Guys, you've got to see this!" Everything stopped while Howie cued up the tape. It only lasted a few minutes, but the memory of what it shows will stay with Tom forever.

Two things from the tape stand out in Tom's mind. First, the tape shows Binti Jua as she ever so gently picks up the boy. It is as if she knows he is injured, and she turns her back on another gorilla, who is walking in her direction, as if to shield him. Second, it shows Binti Jua holding the little boy just as she would a gorilla baby.

Most people know that a gorilla's arms are longer than its legs—the exact opposite of a human's. So the boy's legs are kind of dangling out behind her, but she is cradling him with his head in the crook of her left arm, and his legs and upper body are supported by her right arm. There is no sound on the tape, but Tom can imagine the commotion, and the tape shows Binti Jua looking up in alarm and bewilderment at the visitors above her. This is the part that really gets Tom. As she looks upward in concern and alarm, she is gently and reassuringly patting the little boy's butt with her right hand—just as you might do. And she is holding the little boy just as you would.

The tape did not get into the hands of the media until Saturday night, and when it did, the already-high public interest built even further. The staff held daily press conferences, and by Sunday, film crews from around the world were descending on the zoo. Of course, the zoo got copies of everything. One story, which aired in Japan, cut from Binti Jua to Hillary Rodham Clinton, who happened to be in Chicago that week. Like everyone else, Mrs. Clinton was talking about Binti Jua.

Jay Leno picked up on the story that week, too, on the *Tonight Show.* He did nightly Binti Jua stories for a week, but Tom only remembers one of them. In it, Jay is doing one of his famous fake news stories outside of the studio. The reporter, an actor, is stricken with an apparent heart attack. The ambulance comes, but instead of a paramedic, out steps another actor in a

Binti Jua costume. Binti Jua listens to the man's chest and then gets out the paddles and restarts his heart. Most of the stories, however, were pretty serious in nature and many reached a national audience. The paramedics who assisted in the boy's safe recovery, for example, appeared on the *Oprah Winfrey Show* and, later that year, Binti Jua received *Newsweek*'s "Hero of the Year" award.

They were serious, in part, because the boy could have been killed, either by the fall or, possibly, by an angry or upset gorilla. The staff speculate that it might have been fortunate that there was not a male on display that day, but, several years earlier at the Jersey Zoo, the very famous zoo located off the coast of England, a young boy fell into a gorilla exhibit and was actually guarded by the silverback from the curious females and blackback males until help could arrive. I should quickly point out, however, that this shouldn't be taken as encouragement for anyone to try jumping into a gorilla habitat on purpose.

The news stories were also serious in that they helped people to change their basic attitudes toward gorillas. Many people think of gorillas as ferocious animals. But they aren't, really, unless provoked. They spend most of their lives being exactly what they are—gentle vegetarians that live in fairly peaceful social groups. The fact that Binti Jua displayed compassion and concern helped people to see these animals in a totally different fashion. They identified with Binti Jua in a new and powerful way.

But what amazes me the most is not the effect that the rescue had on people's attitudes toward gorillas. What amazes me is the effect that the rescue had on the other gorillas' attitudes toward Binti Jua. Prior to the rescue, Binti Jua was a low-ranking female. One of the ways you could tell how the gorillas regarded one another at Brookfield was to look at where they spent the most time. Generally speaking, the higher their social ranking, quite literally the higher they were on the mountain. The dominant females tended to monopolize the top of the exhibit, and Binti Jua spent more time lower on the mountain. In fact, that partly explains why it was Binti Jua who got to the child first. Usually she was a lot closer to the exhibit floor than the higher-ranking females. Beginning on Saturday, the day after the fall, people came in record numbers to see the gorilla that had saved the boy. They would point at her and exclaim to one another. They talked about her and admired her. She was the focus of the visitors' attention, and believe me, none of this was lost on the other gorillas. Over the next few weeks, Binti Jua's social status in the group rose, as did she—right up the mountain. While other factors certainly

Binti Jua and her baby. BROOKFIELD ZOO

could have contributed to the fact that Binti Jua spent more time on top of the mountain—for example she might have developed some aversion to being in the moat area because of the commotion of that day—nonetheless there she was, the object of an extraordinary amount of public attention.

When I talk about the tragedy in Philadelphia and the miracle in Chicago, I have to stress that the "miracle" is not simply that the child in Chicago survived his fall. The miracle is that, just as the people of Philadelphia responded, with deep emotion, to the loss of their gorillas, so the gorillas of Chicago responded to the admiring stares of the people of Chicago. Two very different species are thus seen to share richly similar social and emotional lives. It tells us once again that we are not so very different from other animals, and they are not so different from us.

But it is wrong to think that we are the same. Yes, the great apes—chimps, gorillas, bonobos, and orangutans—show many striking similarities to us. But there is also one profound difference. It is has to do with the way we learn.

Clark Hull, a psychologist, tells us that there are three kinds of learning: situational, social, and symbolic. The definitions are simple enough. Situational learning takes place when you have a direct experience with something—for example, when you touch a hot stove. Social learning takes place when you observe someone else's behavior or experience. For example, I can watch you touch a hot stove and, judging by your reaction, can learn not to do so myself. The magic of the third kind of learning, symbolic learning, is that you can, for the most part, be rid of the immediate presence of the original experience entirely. In other words, I can say, "There are really hot stoves in the next room. Watch out!"

Now, clearly, you must have some notion of what those words mean in order to understand what I'm trying to tell you. You must, for example, have some sense of what the words *hot, stove,* and *room* imply. You must also have some knowledge of what linguists call syntax, or word order, and you must have some greater understanding of the context that the words are spoken in. To prove it, let me suggest that if my wife and I were renovating our house (which we recently did) and went shopping for kitchen fixtures in a large store that had several rooms, each filled with different kinds of appliances, and I said to her, "Watch out. The stoves in the next room are really hot," it could be taken to mean that I liked those particular stoves a lot, that she should go take a look at them with me, and that we might wind up spending more money than we had planned.

Animals certainly communicate. Some animals, primates in particular, can be taught to communicate using symbols. But no animals (other than humans) acquire the majority of what they know through language, or symbolic learning.

Let's take a quick look at animal communication. Male fireflies flash a special code, one that is different for each species of firefly. The male flashes his code until he receives a reply from a female, and then they mate. Likewise, bees are able to communicate with one another using a very different kind of code. We have a working beehive in our zoo with glass walls so that visitors can observe the bees in action. If they look closely, what they'll see is worker bees returning to the hive and performing what might best be called a dance. If they dance in a circle, it tells the other bees that there is food relatively close by—within one hundred yards. If they dance in a figure eight, it means that the food will be found much farther away. As they dance their figure eight, they waggle their bodies. The harder they waggle, the further the food source. Now the other bees know how far away the food is, but they don't know which direction to go in order to find it. The dance, however, is performed on the vertical plane of the hive. The angle of the figure eight is the key to communicating the direction of the food source. If they dance upward, as in straight up and down in relation to the wall of the hive, they're telling the other bees to fly straight in the direction of the sun. If they dance downward, the other bees should fly in the opposite direction of the sun and, of course, any angle in between works just as well.

You could say our bees are communicating using symbols (circles and figure eights) in their own language, but their communication is fairly limited and they really can't learn much about the world using symbolic communication except, of course, where food might be found. Primates, on the other hand, seem to be capable of much more powerful communication skills. Washoe, a chimp that has been trained to use the same signs employed by deaf humans who communicate in Ameslan (American Sign Language), *reliably* employs more than two hundred different signs. ("Reliably" means that three different observers have seen Washoe use a given sign in the correct context on each of fifteen consecutive days—a pretty strict definition.) In all, Washoe has used over one thousand signs, but she only uses about one-fifth of them regularly.

Ameslan has also been taught to gorillas, and Koko is probably the most well-known signing gorilla. Like Washoe, Koko invents new terms for objects by combining words that she already knows. For example, she invented the term *elephant baby* to describe a Pinocchio doll, *bottle match* for a cigarette lighter, and *eye hat* for a mask. Koko also communicates some very complex thoughts. For example, when her trainer (Francine Patterson of the California-based Gorilla Foundation) continued to drill Koko on signs for different parts of the gorilla anatomy, she finally signed, "Think eye ear eye nose boring."

Some of the most fascinating experiments in primate communication can be seen at the National Zoo's Think Tank, where researchers are exploring communication skills in orangutans. Think Tank opened in October 1995, and I had a chance to visit the lab, which is open to the public, not long after that. The brainchild of biologists Ben Beck and Rob Shumaker, the Think Tank project picks up where previous studies of language skills and learning in primates leave off. In addition to teaching the animals abstract symbols, in this case using touch-sensitive computer screens and abstract forms instead of Ameslan, the project measures how the language skills can be transferred, or taught, to other orangutans that have not been instructed by human trainers.

Shumaker designed an abstract language that uses what we might call pictograms. For example, the symbol for "banana" is a rectangle with a wavy horizontal line through it. The symbol for "apple" is a rectangle with a straight line and a dot in the center. The symbols, or words if you will, are broken down into seven different categories: food, activities (or verbs), adjectives, nonfood objects, human names, orangutan names, and numbers. Each category of words has a distinct exterior shape—diamonds, circles, rectangles, and so on.

The orangutans have two different computer monitors in their "smart room" enclosure. One of them, called the response monitor, has a touch-sensitive screen that is controlled by the animal, and the other, called the stimulus monitor, is controlled by a researcher. Researchers can ask questions on the stimulus monitor, and the orangutan can respond by touching the appropriate symbol on the response monitor. Since the experimenter and the orangutan can see each other, experimenters aren't necessarily restricted to the use of the computers to communicate. For example, during the sessions I watched, the experimenter had some freshly cut pieces of apple. The animal correctly identified the symbol for apple among the many different words on the touch-sensitive screen, demonstrating that the connection between the actual food and the abstract symbol was clearly understood.

Initially, three pairs or orangutans were involved in the experiment, but only the dominant animal in each pair receives active instruction in the language by zoo researchers. The subordinate animal is left to pick up on the symbols only by watching. In other words, the National Zoo is exploring the relationship between what our psychologist Clark Hull would term social learning, or learning acquired by watching others, and symbolic learning, or learning using abstract symbols.

The Think Tank project is too new to tell us much about how well orangutans can use abstract symbols, or what we would call words, in a concrete way, and it will be years before we begin to understand how effective orangutans can be at learning language on their own or, even more interestingly, learning by teaching one another. In the process we will learn a great deal about how intelligent these animals really are. There is no doubt in my mind that this project will demonstrate that orangutans can acquire language skills, express thoughts and emotions, and, most likely, teach each other.

They can certainly teach us. My favorite story on orangutans as teachers comes from Dr. Lee Simmons, the longtime director of the Henry Doorly Zoo in Omaha and one of the nation's premier zoo administrators. Lee worked for many years around an orangutan named Fu Manchu, who we'll meet again later in this book. Fu had come into zoos at a relatively young age. Although he was considered relatively safe to be around, as an adult male, Fu could be unpredictable. Still, keepers routinely went into his enclosure, a testament to Fu's gentle personality. One day, Lee and some of his staff were in Fu's exhibit when Lee's senior curator slipped on the wet floor and instinctively reached out his hand to brace himself against the exhibit glass. The problem was, the window was coated with an electrically charged film, designed to keep the orangs from finger painting the window with dung. Lee, who really is a sensitive soul at heart, doubled over with laughter. The curator, on the other hand, wound up flat on his back with an embarrassed and slightly dazed look clouding his face. Fu watched the whole episode with obvious interest. As the curator lay on the ground, glaring at his fellow employees, Fu ambled over, gently took the curator's hand, held the man's forefinger out, and touched that finger to the inch-wide strip of glass around the frame that was not electrified. He seemed to be saying, as Lee put it later, "Look, you dummy. If you touch it *here,* you won't get knocked on your rear!"

So, it's clear that chimps, gorillas, and orangutans share, with humans, strikingly similar abilities to teach and learn. On the other hand, the differences between humans and great apes in terms of the *extent* of those abilities is almost unimaginably vast. To prove it, I will make you this bet. I bet you that no matter how tirelessly researchers and scientists work with great apes, they will never be successful in getting one of them to understand any single sentence, selected at random, from the *Encyclopaedia Britannica* (unless there's one in there that says, "You give cup raisins"). On the other hand, there is not a single sentence in the *Encyclopaedia Britannica* that the average human cannot be made to understand. In fact, if you're reading this book, my guess is that there is not a single sentence in the *Encyclopaedia Britannica* that you couldn't understand already!

The latest edition of that encyclopedia is sold in a bound set of thirty-two volumes. That is a huge amount of information, most of it in the form of words. By comparison, the New York Public Library had 14,310,969 books, last time they counted. That is but one library in one country. Add to that the amount of information in all of the books in all of the languages in all of the libraries in all the world. Add to that the amount of information in all of the newspapers and magazines, plus the information broadcast daily via radio and television and other media. Add to that all of the conversations that we have between all of the people on our planet every day. Now, compare that to the sum total of what a chimp, gorilla, or orangutan can access via symbolic learning.

The capacity for symbolic learning sets humans apart. Yes, many animals can communicate using symbols. But all of those forms of symbolic communication are strictly limited. In the case of humans, we can communicate so much more that a quantitative difference becomes a qualitative one.

Now I could read the *Encyclopaedia Britannica,* understand what I read, and not necessarily remember it all. Our family, when I was growing up, was more of a *World Book* family (I guess that's some sort of index of social class), and I did, in fact, read the encyclopedia for the fun of it. Not cover to cover, mind you, but I would pick a volume at random and just start reading. I don't think that I, or anyone else for that matter, could retain all of the information, though.

The interesting thing about information is that while *no one* could know all of the things there are to know, together among us, all of us know what is known. The capacity for symbolic learning makes it possible for us to acquire some small part of all that is known (small, that is, relative to the amount that there is to know, but huge relative to what any other species could ever acquire). More importantly, it makes it possible for us to create knowledge. The ancient Greeks, for example, defined a good conversation as one wherein two people, in talking to one another, learn something that neither of them knew before. Symbolic learning sets us apart from all other species.

The difference between humans and other primates ultimately comes down to our capacity for symbolic learning. Apes learn a great deal about the world through situational learning. They are also great social learners. They also have a limited capacity for symbolic learning, and this capacity is certainly worth investigating. But only humans are great symbolic learners, and the amount of information that we as a species have amassed and can access makes us absolutely different from any other species ever.

We share much in common with other animals, particularly with other primates, and especially with the great apes, but our capacity for symbolic learning sets us absolutely and irrevocably apart. Yes, we share similar emotions. Yes, we form similar social groups based around families. Yes, we are all intelligent and sensitive. And yes, we can learn by direct experience, by watching others, and by using language. But the human capacity for symbolic learning is *not*, as Mark Twain might have it, the difference between lightning and a lightning bug. It is far, far greater. If it weren't, the movie *Planet of the Apes* would be reality, not the work of our powerful imaginations.

4

When a Butterfly Sneezes

It's funny how the smallest things can have a huge impact on our lives. My favorite example is the story of how I *really* graduated from college. I spent my senior year working on an honors thesis in the anthropology department at the University of Missouri. I wrote a very long paper on the evolution of law in society—probably pretty boring stuff, but I had poured my heart into it. The catch was that the grade for the thesis (and the corresponding six hours of college credit—more than enough to keep me from graduating if I didn't have them) was based, not on the thesis, but on my defense of the thesis. In other words, I would have to sit down with three professors from the anthropology department and defend my ideas in the face of their verbal barrage.

I had heard stories of how awful this grilling could be, and I was more than a little frightened at the prospect. But I really trusted my main advisor to gentle me along. Misplaced trust, as it turns out, can be the bane of our existence.

My thesis advisor had an office in the front of Switzler Hall, a beautiful old building from the 1800s. The office had high ceilings, two tall windows that looked out on the university's famous quadrangle, and walls lined with anthropology books, lovingly arranged by author. The old, yellowed shades seemed to be always pulled most of the way down, and the office light was warm, but subdued.

The four of us met on a beautiful morning of blue skies punctuated by bil-

56

lowing white clouds. I remember thinking that a pretty good day was about to be ruined by a pretty bad meeting. But I wasn't too worried. My advisor was going to ask the first questions, and I figured he'd toss me some softballs to get me warmed up and to impress the other two questioners.

So we sat down, and my advisor asked the first question. He said, "Can you name three scholars of anthropological jurisprudence from the nineteenth century who were not anthropologists by training, but who still made substantial contributions to the anthropology of law?" I about puked. Not only could I not do that, but I don't know of anyone else who could. That question couldn't be answered—not in a million years. I was in big trouble.

I later learned that this is a normal part of this peculiar rite of passage. They always make it tough on you to join the club. That's why, in our culture, we have initiations. Anyway, I was stumped, but not completely. I had recently read a book by a man named Morton Fried, who was later to guide me through a similar defense, this time for my Ph.D. In his book he talked about a scholar named John Austin, a nineteenth-century lawyer who had thought and written a lot about law and politics. All I could remember was that Professor Fried had referred to Austin as "that crusty proponent of positive polity." I had no idea what that meant, but it struck me as very funny. (Academics are like that.) Anyway, I confidently said, "Well, there's John Austin, whom Morton Fried refers to as 'that crusty proponent of positive polity,'" and we all had a good long chuckle.

While they chuckled, I began to pray. Not in earnest, because there obviously wasn't much time—but I was fervent nonetheless (proving once again that there's no such thing as an atheist in a foxhole). Just then, one of those puffy white clouds moved just enough for the sun to shine bright again. A ray of sun broke between the edge of the shade and the window's edge, casting a narrow band of sunlight down the spines of those carefully arrayed books. Two books, side by side, caught my eye. One was by Maine and the other by Morgan. I didn't stop to think if I was right or not; I simply blurted out, "and, of course, there's Maine and Morgan." Actually, to this day, I don't know if any of those three names would qualify as correct answers to the question. But it doesn't matter. They accepted my answer. Furthermore, they realized that since *they* couldn't possibly have answered that question, they were likely to expose their own relative ignorance by pursuing the topic. I wasn't supposed to know the answer, but it seemed to them as if I did. So after a pause, they asked me when I was leaving for graduate school and if I was having a hard time finding an apartment in Manhattan.

And so the tiny movement of a cloud changes one young man's day and, perhaps, his life.

If small things can result in great changes, then how much more so great events? Or terrible events?

As an example, I trace the fact that I am sitting here in this office today, writing these words, directly to the actions of a small group of terrorists thousands of miles away.

On the morning of September 11, 2001, I was in St. Louis for the American Zoo and Aquarium Association's annual meeting. It was, until that morning, a particularly good meeting. The Saint Louis Zoo looked wonderful, and I enjoyed walking around the grounds. It had certainly changed enormously from the days when I had enjoyed my hot dog–and–beer lunches near the sea lion basin, and all of the changes were for the better.

I was anxious to get back to Indianapolis, though, and after going for a morning swim in the deserted pool on the roof of the downtown Adam's Mark hotel, I returned to my room to pack my bags for the four-hour drive back to Indianapolis. As I stepped onto the elevator, a few of my colleagues noticed my bags and made some comments about flying back. I can't remember what I said, but I know my remarks were somehow inappropriate because everyone looked at me oddly. At that point, apparently everyone in the hotel had heard of the tragic events of that morning except, it seemed, me. I've tried to remember who it was on the elevator and what I said, but I can't recall. Anyway, if you're reading this and you were on that elevator on the morning of 9/11, I apologize for what must have seemed to be some pretty inane comments.

When I stepped off the elevator in the main lobby, everyone looked harried, and half of the delegates seemed to be talking on cell phones. "This conference," I thought to myself, "has gone on too long for most of these guys." Then my own cell phone rang. My assistant was calling to find out if we should immediately close the zoo and send all unessential personnel home. "What are you talking about?" I asked, "Why would we do that?" And that is how I learned about the World Trade Center.

Had it not been for that tragic day, I might never have wound up leaving Indianapolis for St. Louis. No one, of course, could get back to their zoos or businesses by air, and the staff of the Saint Louis Zoo did everything in their power to get people home. They arranged for rental cars, chartered buses, and invited people to stay in the homes of staff members and volunteers. One of the individuals stuck in St. Louis was James Abruzzo, who ran an executive recruitment firm—he was a headhunter. He had been in St. Louis to meet with the commissioners of the zoo, who had just hired him to find a replacement for Charlie Hoessle, who had headed the zoo for eighteen glorious years and, at the age of sixty-nine, had just announced his retirement.

James was lucky enough to find a rental car by the morning of the thirteenth, and he began the long drive back to New Jersey. His route took him past Indianapolis, and he decided to stop by and visit with me to see if I had any thoughts about who would make a good replacement for Charlie.

As it happened, I wasn't the least bit busy when he showed up unannounced at my office to ask if I might have any time to see him. I don't how well you recall the first few days after 9/11, but really, people weren't venturing out of their homes. The zoo was deserted and, if I had any appointments, they had all been cancelled. (Not surprisingly though, by the next week people were coming to the zoo in record numbers. Perhaps they wanted to be in a safe, secure place; perhaps they wanted to be surrounded by the reassuring presence of life; or perhaps they wanted to do something that was family-friendly. For whatever reason, every zoo in the nation reported a similar surge in attendance.)

So I met with James. He knew, I think, that I had once worked in St. Louis, but he didn't know that my family was from there and that my wife still had literally all of her family living in St. Louis. I don't think he thought that I would ever leave Indianapolis. I know I didn't think I ever would. But the longer we talked, the more interested I became. After an hour or so, I was thinking the unthinkable. I was thinking about going back to St. Louis. Had it not been for the tragic events of 9/11, James would never have stopped in to see me, and I doubt I would ever have left Indianapolis.

My favorite example, though, of the fortuitous nature of small events as they cause enormous change, comes from the island of Jamaica. Prior to the arrival of the earliest European colonists, the largest native land animal on the island was the Jamaican ground iguana, *Cyclura collei*. Stretching upwards of five feet when fully grown, the ground iguana inhabited the southeastern part of the island, roaming the dry limestone forests. They were hunted for food by the native Arawak Indians and by the early colonists, but they held their own in terms of numbers until 1872. In that year a predator was introduced to their island that was even more deadly than human hunters—the Indian mongoose.

With the arrival of the mongoose, the iguana's numbers plummeted, and by

1900, the only iguanas thought to be left alive were on Goat Island, a tiny island off the south coast. Of course, it was only a matter of time before the mongoose could hitch a ride to the island, and by 1920, the end of the ground iguana was near. The mongoose invaded, the population fell, and biologists were forced to mount a major rescue effort. In 1940, they captured the twenty-two remaining iguanas and attempted, without success, to breed them. By 1946, all of the captured iguanas were dead, none had reproduced, and the species was certified as extinct.

But like a phoenix, they were to rise from the dead. Over the next thirty years, there continued to be sporadic reports of huge lizards seen roaming the Hellshire Hills, the ruggedly wooded peninsula on the mainland across from Goat Island. There were never any confirmed sightings, but the rumors eventually got back to a scientist named Jeremy Woodley at the University of the West Indies.

Professor Woodley made occasional trips to the Hellshire Hills to interview the people most likely to come across the huge lizards—charcoal burners, who go into the forest to cut down trees to make into charcoal, and pig hunters, who hunt the wild pigs that were also introduced to the island by the colonials. In 1970, his persistence paid off. He found a dead iguana, the first confirmed sighting on the island of Jamaica in seventy years. So they were known to still exist, but where and in what numbers remained a complete mystery. The ground iguana came to be known as possibly the most endangered species of lizard in the world. Indeed, for the next twenty years, we were never completely certain that Woodley's dead iguana wasn't the very last one of its kind.

Fast-forward twenty years. A man named Edwin Duffus takes his dog and his gun and goes into the hills in search of wild pigs. His dog comes out of the scrubby trees dragging an enormous lizard. Mr. Duffus captures it—alive.

This is the part that amazes me. He could have killed it. I know many people who would have killed the giant reptile without a thought. If I thought the struggling monster was about to injure my dog, perhaps I would do the same. But he didn't. He could have gone home with no one the wiser. Perhaps he would have told his tale around the village where he lived, and perhaps his sighting would have become another in a long line of rumors related to the iguana's persistence. But no, Mr. Duffus took his lizard immediately to the Hope Zoo in Kingston, the island's capital. And this begins the story of the Jamaican ground iguana's rise from oblivion.

The Hope Zoo got together with the Jamaican Natural Resources and Conservation Authority, the Institute of Jamaica, and the University of the West Indies to form what they elected to call the Jamaican Iguana Research and Conservation group. The group almost immediately decided that they needed a well-trained and highly motivated field research technician to begin a systematic search for the ground iguana. They hired Mr. Duffus. Okay, he was a pig hunter—but he found the damn iguana, and if he could find one, so the logic went, he could find more.

And he did. He led a team into the hills and, over the next six months, they posted twenty-three sightings of at least fifteen different iguanas. Once they knew what to look for, evidence of the iguana was everywhere. Fecal pellets, shed skins, and the distinctive marks of a very long tail dragging across the ground convinced them of the iguanas' presence. Evidence of two small nesting sites convinced them that they continued to breed.

The Hellshire Hills are a difficult place to work, but as it turns out, they are an important place. The terrain is covered with razor-sharp rocks, and the topography is steep, but the environment is very rich from the point of view of a conservation biologist. The forest features over three hundred species of higher plants, including fifty-three species that are found on Jamaica and nowhere else in the world. There are reptiles and birds in this forest that cannot be found elsewhere. The remoteness and inaccessibility of the Hellshire Hills has made this dry upland forest one of the last of its kind anywhere in the West Indies.

With the Hellshire Hills now being the focus of a concerted effort to find and save the rarest lizard in the world, the Hope Zoo found itself in the spotlight as well. It was not really a well-known place in Kingston, let alone in Jamaica. It was, like many zoos in that part of the world, small, poorly funded, and understaffed, yet it was now expected to launch a conservation effort that was international in scope and importance. But as is so often the case, the Hope Zoo made up for in talent what they lacked in resources. They had an extraordinarily dedicated curator, named Rhema Kerr, who took it upon herself to contact a small group of American zoos and forge a remarkable partnership.

Ms. Kerr contacted the Fort Worth Zoo and the Indianapolis Zoo, and they, along with the Bermuda Zoological Society, funded a meeting of a small group of government officials, conservation and field biologists, and experts from American zoos. The meeting was orchestrated by CBSG, the Conservation Breeding Specialist Group. CBSG was the brainchild of Ulie Seal, the same scientist who was responsible for organizing the program to save the partula snail. CBSG is a specialist group of the International Union for the Conservation of Nature and Natural Resources (IUCN). The IUCN is the international body of scientists and conservationists that administers CITES, the treaty that regulates and restricts the trade of threatened and endangered species. Most of the specialist groups in the IUCN deal with species of plants and animals, but some of them, like the group that works with translocating wild animals from one place to another, or like CBSG, which helps us to understand how zoos can use captive breeding to augment or reintroduce animals to the wild, deal more with managing wild animals than studying them.

When trying to figure out what to do to save a species, the members of CBSG look at two things. First, they want to analyze where the population is headed and what the continuing threats to its survival might be. Second, they want to get a buy-in on solutions from every possible stakeholder. It simply doesn't pay to figure out how to save a species if the plan can never be implemented.

What emerged after three days was a simple prognosis. Mr. Duffus had not found many juvenile iguanas. When they plugged the key assumptions into the computer—first, that the overall population could be as low as fifty individuals and as high as two hundred, and second, that most of the individuals were older—they concluded that the chances for survival were very low. But a plan was hatched, so to speak, that could save the remaining animals.

At the core of the plan was the idea that if the baby iguanas could survive long enough, they'd be too big to be killed by the mongooses. Enter, again, the irrepressible Mr. Duffus. He was drafted to find the breeding sites, protect the nests until the babies were born, and then take the babies to the Hope Zoo to grow up. After five or six years, the thinking went, they'd be big enough to be returned to the wild and could survive on their own. In other words, they came up with a "head-start program" for lizards.

Next, in 1994, Ms. Kerr sent twelve of the captive-raised iguanas to the United States. In 1996, she sent another twelve. Some went to the Fort Worth Zoo, some went to the Indianapolis Zoo. Ms. Kerr's logic in sending animals out of the country was pretty simple. First, she figured that if a major catastrophe occurred in the wild population, say, the introduction of a disease, she would have a safety net. Second, little was known about how to breed the ground iguana in zoos. Remember that the last time it was tried, all of the animals died without ever reproducing. It was, and still is, essential to have as many people as possible trying to crack the riddle of their reproduction.

Breeding reptiles is part science and part art. You would think it should be *all* science, but it's not. We can't really do it as scientifically as we would like because we usually don't have enough animals to start with to have things like experimental controls. What we do, instead, is take what we know about the natural history of an animal and then begin varying certain conditions while hoping that something will work. For example, we might see what plants or fruits come into season when they breed in the wild. It could be that diet triggers breeding. Or it could be that they breed after they hibernate. We'll try popping them into the refrigerator to simulate hibernation, and see what happens when they come out. It could be that males need to be in competition with one another to cause them to breed. So we try introducing multiple males to a female at breeding time. Light levels, territory size, and a host of other factors could come into play. We simply keep trying until we hit on the right combination.

Of course, captive breeding is only a part of the long-term solution. The Hope Zoo has undertaken a number of other programs, both in the zoo and in the wild. In the zoo they (and their partners in the United States) are studying juvenile feeding patterns (a study that the Saint Louis Zoo participated in, along with the Fort Worth Zoo and the University of the West Indies), how the iguanas interact socially, how their bodies respond to changes in temperature, and how their activity patterns vary through the day. In the field, the conservationists began trapping and removing mongoose to decrease predation, and they undertook a radio-tracking program designed to monitor the newly released "head-start" animals. I remember marveling that, after researchers had tried numerous high-tech materials for the vests that the iguanas could wear in order to carry their radio transmitters around in comfort, the fabric that worked the best was good, old-fashioned denim.

Perhaps the most important field programs, however, have nothing to do with iguana research at all. In the long run, their little tiny breeding area in the Hellshire Hills, an area not much larger than this floor of my building, in a forest that is only some six miles by ten miles in size, must be protected from hunters and their dogs and from the charcoal burners who push deeper into the woods every year in search of wood. It is here that Mr. Duffus shines the most. Since he joined the team, he has worked tirelessly to help convince the people of the region that the iguana is worth saving. His efforts seem to be working, although this precious little habitat is still threatened.

When I last checked on the Jamaican ground iguana, late in 1998, twenty head-start animals had been released. We still have not figured out how to breed them in zoos, but every season we try something new. Eventually we will unlock the mystery. But in the meantime, I think about the extraordinary good fortune that brought Mr. Duffus to meet the Jamaican ground iguana. Had it not been for this remote chance encounter, with this particularly remarkable man, the iguana probably would have faded away forever.

When I lived in India, my friends had an expression that translated roughly like this: "When a butterfly sneezes, the world trembles!" That's an elegant way of saying that everything is related to everything else—that things are interconnected, and even small things, like a chance encounter on a pig-hunting trip, can have enormous results.

That's one of the ideas that we must always strive to understand if we are ever to understand the world around us. All things on our planet are related to one another. Even the smallest of actions, to one degree or another, affect this complex web. Irrespective of magnitude, events ripple around our world in ways great and small. Scientists understand this well, but they are often at a loss to document it completely in nature. The complexity of the interconnections are too vast, the ripples too extensive, to ever fully comprehend. Says E. O. Wilson:

> Eliminate just one kind of tree in hundreds in a forest, and some of its pollinators, leafcutters, and woodborers will disappear with it, then various of their parasites and key predators, and perhaps a species of bat or bird that depends on its fruit—and when will the reverberations end? Perhaps not until a large part of the diversity of the forest collapses like an arch crumbling when its keystone is pulled away. More likely the effects will remain local, ending with a minor shift in the overall pattern of abundance among the numerous surviving species. In either case the effects are beyond the power of present-day ecologists to predict. It is enough to work on the assumption that all of the details matter in the end, in some unknown but vital way.

As difficult as it is to comprehend this web in nature, it is even more difficult to understand when humans are added to the mix. Science recognizes three great realms in the world as we currently understand it. First, there is the realm of the *inorganic*. This is the realm of nonliving things, the elements—like carbon and iron and hydrogen and oxygen—that combine and recombine to make up everything around us. We borrow from these elements to make living things. I, for example, am mostly made of water, a combination of hydrogen and oxygen, and like most life forms on our planet, there is a good deal of carbon in me. I need oxygen to power my body, and so I use atoms of iron to bind the oxygen and carry it to my tissues. Of course, I need many other elements, in combination, to sustain life, but oxygen occurs to me first. I can go without food for long periods of time, water for much less time, but I can only go without oxygen for a matter of a few minutes before I would die. Living things, like me, make up the second great realm of our world, the *organic*.

The third great realm, after the inorganic and organic, is the one most difficult for us to comprehend. The third great realm is what social scientists refer to as the *superorganic*. Just as the realm of the organic is somehow greater than the sum of the parts of the inorganic pieces of matter that compose it, the realm of the superorganic is greater than the sum total of the organic things that, in turn, compose it. The superorganic can be roughly equated to humanity. Think of it this way—we are all individual organic entities, but we combine in complex social groups. These groups have distinctive patterns of acting as well as distinctive patterns of thought and feeling. Anthropologists refer to these patterns as our *culture*. Culture is, quite simply, everything about us. It is how we use the natural environment to fuel our bodies, on an individual basis, and our economies, on a social basis. It is how we organize ourselves to continue our species in families, again on an individual basis, and larger kindred groups, communities, and nations, on a social basis. It is how and what we teach our children. It is how we govern ourselves and direct the greater social system. It is our values, beliefs, and attitudes, our philosophies and our ethics, our art and our literature. And it makes humans special. We are the only species that has it. Although gorillas have brains that can do things very much like the brains of humans, still, the closest gorillas have ever come to having culture was in the movie *Planet of the Apes*. Only humans have culture.

Culture rests, ultimately, on our ability to communicate with each other using symbols and, as we have seen, to use our capacity for speech as the basis for how we learn. It makes for differences that are both subtle and profound. I am reminded of a silly quote that might capture the distinction. "After all, in what consists the difference between man and beast, save in the former to learn the art of dining, while the latter is forever doomed to feed?" Culture transforms feeding into dining. The difference has nothing to do with taking in organic compounds (food) seasoned with inorganic compounds (like salt, a combination of sodium and chlorine). It has everything to do with culture. Says E. O. Wilson, "Culture in turn is a product of the mind, which can be interpreted as an image-making machine that recreates the outside world through symbols arranged into maps and stories."

All this is important because the superorganic has great power to redirect the organic, just as the organic can organize the inorganic realm into the amazing and complex world of life. Understanding this process, though, and channeling it toward a specific end, is the thing we have never fully mastered.

One last example from Professor Wilson will help to illustrate this. He says that there is at least one thing that virtually all politicians, diplomats, soldiers, and citizens can agree on. We are convinced within our culture and within virtually all of the other cultures of this planet that the worst thing that could befall our world is a global nuclear war. We have the capacity to wage such a war, to wreak wholesale destruction on our planet, to make the earth uninhabitable for ourselves and every other living thing. We understand it and we as a culture of cultures do everything in our power to ensure that such a thing never happens. Now:

> With that terrible truism acknowledged, it must be added that if no country pulls the trigger the worst thing that will *probably* happen—in fact is already well underway—is not energy depletion, economic collapse, conventional war, or even the expansion of totalitarian governments. As tragic as these catastrophes would be for us, they can be repaired within a few generations. The one process now going on that will take millions of years to correct is the loss of genetic and species diversity by the destruction of natural habitats. This is the folly our descendents are least likely to forgive us.

So the question remains, "How do we affect our culture to halt this terrible destruction?"

Here is where the realm of the superorganic comes to the fore. You see, it is not enough that I do not wish this to go on happening. It is not enough that you and I do not wish it. There must be a great many of us who hold this value, this belief, this attitude, before our culture, and thus our behavior, will change. Yet, magically, it starts with just a few of us. We know it can be done, but because we are very small pieces in the midst of something very large, individual pieces of a vast superorganic realm, it is hard for us to see how. When it comes to fundamentally changing our relationship with the natural environment, it can truly be said that we cannot see the forest for the trees. Still, I say again, it can be done. Even though this great change in our cultural system must occur in ways unfathomable to ourselves, we know the change can begin in something very small.

Perhaps it can begin in something as small as a butterfly's sneeze.

5

The Telemetric Egg

The Saint Louis Zoo has the largest collection of waterfowl of any zoo in America—at last count we had 442 individuals representing 61 different species. They live, for the most part, on the three small lakes in the center of the zoo. The problem is that the birds in our collection face many of the same difficulties that birds in the wild face when trying to raise their young, the biggest being the threat of predation. We're located in Forest Park, and the park is full of wild animals—raccoons, fox, and others—that can and do feast on our eggs or our baby birds.

In order to understand this threat, we're currently working with scientists at Washington University and the University of Missouri–Columbia to study the predation. The first part of our study is focusing on raccoons. We set traps for the animals, catch them, and then put radio collars on them. The collars help us to understand their movements, which helps us to understand what they're feeding on and where they're feeding.

While it's important to understand predation, it's even more important to keep our collection safe. The best way to protect eggs and very young birds is to take the eggs out of the nest, hatch them someplace safe, and raise the baby birds until they're old enough to fend for themselves. The problem is that birds are much better at incubating eggs, at least for the first ten to fourteen days after the eggs are laid, than we are. They're better at it because they instinctively know the ideal temperature for the eggs and, just as importantly, how often the eggs should be turned over in the nest, at various stages in the eggs' incubation.

We could, I suppose, chase the hen off the nest, mark the eggs and take their temperature, and then repeat the process every day, several times a day, for two weeks. The problem with that is obvious. We probably wouldn't get very good data, and the procedure would probably result in some very irate ducks.

Our solution to the problem has been to develop a telemetric egg—a fake egg with a sensor inside that measures temperature and turning rate—coupled with a miniature radio transmitter that sends a steady stream of data to a computer. Our keepers watch the birds closely and check nests in order to determine when the clutch is complete—in other words, when all of the eggs have been laid and the hen is ready to begin incubating. On this day, one egg is removed from the clutch and replaced with our special, high-tech egg. We then allow the hen to incubate naturally for ten to fourteen days and then remove all of the eggs (including the telemetric egg) and place them in an incubator.

The hen, finding all of her eggs gone, will almost always respond by laying

Telemetric eggs from the Saint Louis Zoo's research department. SAINT LOUIS ZOO

a second clutch within a few weeks. Again, we sneak one out and replace it with a telemetric, but this time we let the bird naturally incubate the eggs until they all hatch. This allows us to make systematic comparisons between artificial and natural rearing techniques using data from the same hen in the same breeding season. It also, of course, allows us to breed twice as many birds.

Telemetry can be used for other things. We have implanted sensors in alligators to see how they thermoregulate (raise or lower their body temperature) and in a variety of mammals in an effort to measure temperature changes associated with ovulation. On top of that, there are numerous applications to field studies. For example, our curator of herpetology (our snake guy) is a man named Jeff Ettling. Jeff's passion is mountain vipers, a group of eight species of venomous snakes found primarily in Asia Minor. The species at highest risk are found in the Caucasus Mountains of Armenia.

The problems with Armenian vipers are, first, that they're vipers (and people who live with vipers tend not to like them), and second, that they're beautiful vipers (and people who collect vipers like the pretty ones the best). This means that they're persecuted by local people, who fear them, and collected intensively by people who love them—largely for the European pet trade. We don't know much about them, due in part to their restricted distributions and the fact that they live in isolated, rocky habitats.

Jeff is studying them by implanting little transmitters under their skin. The signal doesn't go very far, but it is strong enough to allow Jeff and his colleagues to sweep a small radio receiver back and forth over the terrain, listening for each snake's signal. Eventually, they will use the data they gather to better understand the size of each snake's home range, its seasonal activity patterns, and its specific habitat preferences.

The trick is getting the transmitter implanted. Jeff is not a vet, but he trained himself how to slice deftly through the snake's skin, insert the small transmitter, and sew up the incision. As you can imagine, this is a little tricky with a venomous snake. The way he does it is to slide the snake into a large Plexiglas tube, and then he gently feeds the snake into progressively smaller tubes until he gets it to the point where the snake fits pretty snugly and can't wiggle around. The last tube has an opening where he wants the transmitter to go, so Jeff can administer a local anesthesia, perform his surgery, stitch up the animal, and release it. While he does the surgery, he also collects blood samples, partly so we can develop a medical database and partly so that we can perform a genetic analysis designed to tell us how different the vipers are from one another.

A while ago I signed another purchase order from Jeff. He wanted a horse. I guess his vipers are traveling a little farther than he thought, and it isn't easy getting around the mountains of Armenia in a vehicle.

For larger animals, we can employ very different types of tracking technology. For example, in parts of Kenya, elephants are now wearing cell phones. The cell phones allow scientists to track the movements of the elephants, ultimately making it easier to predict and respond to situations where elephants wander into areas where humans live. This is critically important, because an elephant can cause enormous damage to a farmer's crops in a very short amount of time. Reducing human/elephant conflict is a prerequisite to keeping Kenya's 16,000 elephants safe. (By the way, forty years ago, Kenya had 170,000 elephants—so keeping the remaining ones alive becomes more important by the decade.)

One of our partners with our Grevy's zebra project, also located in Kenya, is Professor Dan Rubenstein of Princeton University. Dan has taken the whole cell phone thing one giant step further by having the zebras "call" each other as well as periodically calling home.

The effort started when Dan noticed that the electrical engineering department at Princeton was working to develop campus tours for undergraduates using a global positioning system to give the students information about the campus depending on where they walked. Dan immediately realized that the same system could be used to record the location of zebras and, coupled with some other electronic gadgets, could also be used to record whether the zebras were eating, moving, or resting.

The added twist is that the phones periodically broadcast a signal to search for other collars on other zebras. Once two collars establish a connection, they download to each other. So with each swap of data, the collars not only capture information about the two animals, but also store information about every other animal that *both* of the animals have been in contact with. So now, we don't have to relocate every animal. We only have to find a few animals with collars, and we can download information on pretty much the entire herd. In other words, Dan can now gather data on many animals, the data is redundant (hence more reliable), and most important, we don't have to be watching individual animals to record what they're doing.

Of course, Dan is also thinking more broadly. "What if we put collars on predators like lions?" he asks. If we do, we can begin to understand predator/prey dynamics much better. For example, we might be able to discover if the prey are pulling the predators around the landscape versus the predators pushing the prey. It's interesting to note that the payoffs for developing such systems go far beyond helping to understand animal ecology. Margaret Martonosi, the electrical engineer at Princeton who is working with Dan, says that "remote sensing networks," like this one, could ultimately be used for things like unmanned military surveillance, monitoring underground water pollution, or tracking the weather. There are many other potential applications that could, in the end, make our lives better.

Many of these radio telemetry devices are actually developed in zoos for use in the field. One of the Saint Louis Zoo's scientists, Dr. Karen DeMatteo, has been studying bush dogs in the Amazon basin. Her problem was developing a collar that would stay on the little guys—they have short, thick necks that allow collars to slide easily over their heads; also, they are social animals that, like domestic dogs, like to chew. I can attest to the problem personally. (We have two English bulldogs that love to chew each other's plastic collar clasps. We went through about ten invisible fence collars before we finally had to buy collars with steel clasps and have their battery-powered chargers specially attached to them.) Karen developed her new collars here at the zoo and is now down in the Amazon trying to figure out how to find the dogs. She's developed two techniques that work pretty well. First, she broadcasts their vocalizations to attract the dogs toward her study area; then, she uses special scents to draw the dogs into traps so she can immobilize them, collar them, and release them for tracking.

Technology has changed zoos in countless ways, especially over the last decade or so. Zoo nutrition is a classic case in point. In the early years, nutrition programs for different species were developed either by using diets that we knew worked with domestic animals on the assumption that they should work pretty well with their closely related wild counterparts, or by simply looking at what they ate in the wild and trying to come up with something that was close enough. After all, the logic went, grass is grass if you're a cow (of some sort).

We now know that even slight changes in the nutritional composition of an animal's diet can have a profound effect on its health, behavior, and reproductive capacity. The woman that is in charge of animal diets here at the zoo, Dr. Ellen Dierenfeld, has spent the last twenty years working to improve diets in a more scientific fashion. For example, in order to analyze both what animals eat in the wild and what we feed them here at the zoo, she uses a technology called near infrared spectroscopy. We're all familiar with visible light, but there are also waves that we can't see. At one end of this spectrum of waves are short, high-energy waves like X-rays. At the other end are long waves like radio waves. In the near-infrared portion of the spectrum are the longer waves just beyond the level that is visible. Using this part of the spectrum, researchers can "see" the organic bonds that make up proteins, fat, fiber, and moisture. Ellen can analyze the grass (or anything else) that animals eat in the wild and then develop diets that closely mimic their wild fare.

Ellen has gone one step further and created a computerized database, now housed here at the zoo, that lets dieticians all over the world have access to all that we know about the dietary needs of every species in our care coupled with all that we know about the nutritional composition of different foods. Every year we learn more, and every year she updates the database.

Satellite technology has now progressed to the point where we can actually analyze vegetation from outer space, although not with anything like the precision that Ellen can obtain by analyzing specific samples collected on the ground. For example, the Saint Louis Zoo recently funded a study, by Dr. Glenn Green of Indiana University and Dr. Robert Sussman of Washington University, that utilized satellite-mapping techniques to analyze the vegetation of southern Madagascar. We then correlated the data with the population density of ring-tailed lemurs in the same area. This has helped us to understand what type of habitats lemurs are currently found in and, more importantly, helped us to conduct much more accurate census work.

But my favorite story about the use of satellite technology in conservation doesn't come from the Saint Louis Zoo—it comes from the Bronx Zoo. On October 27, 2001, researchers at the Wildlife Conservation Society in New York thought for a short time that they had made the most important discovery ever. They were looking at satellite images, taken at night, of the remotest parts of our planet. They were interested only in nighttime images because they were looking for lights or, more accurately, for places without lights. Their job was to find areas that would make good wildlife preserves, and their logic was simple: no lights, no people, and where there are no people, there is the potential for wild things to live in peace.

They were looking at images of the area near Patagonia, at the tip of South America, and what they saw absolutely flabbergasted them. They saw, in this isolated and mostly uninhabited region of the world, the lights of a great city—a city even bigger than Sao Paulo, bigger than Buenos Aires. Even more shocking, the city was clearly not on land. It was 175 miles out in the Atlantic Ocean.

They must have immediately thought the same thing that you or I would have thought: "We've found Atlantis! Someone in that ancient and hidden civilization, for one night, must have forgotten to turn out the lights."

The paradise of Atlantis, revealed in the fraction of a second it took to snap a picture in space and send it to earth, was lost over the next few days. When they investigated the image, they discovered not a city under the sea, but instead an armada of fishing boats. The boats were using enormous lights to attract squid to their purse seines, finely meshed nets that they use to haul the squid (and anything else within their grasp) from the sea.

You see, those fishermen are doing what scientists call "fishing down the food chain." Fishermen used to fish for large fish like swordfish—the big fish that feed on the smaller ones. As they exhausted the worldwide stocks of large fish, they began to fish for smaller and smaller fish, exhausting those stocks in turn. The armada off the coast of South America had already taken so many hake that it was no longer possible to fish for them. In fact, they had stopped fishing for fish entirely. They were only keeping the squid. Observers estimate that the boats only keep about 60 percent of what their great nets haul in. The rest is called "bycatch" and is wasted. Animals like seals, sea lions, and penguins are drowned in their nets, and their bodies are simply thrown back into the sea.

The Food and Agriculture Organization (FAO) of the United Nations estimates that 60 percent of the world's fish stocks are "in urgent need of management" to rehabilitate them or keep them from being wiped out entirely. Many regions of the North Atlantic, once teeming with cod, haddock, and bluefin tuna, are now largely barren. A problem that, prior to the 1950s, was confined to a few areas like the North Atlantic, North Pacific, and the Mediterranean Sea is now spreading around the globe. Fish are the last wild creatures to be hunted on such a grand scale, and our fishing technology is now so sophisticated that we can wipe out vast stretches of ocean in remarkably short periods of time.

We should have learned our lesson much earlier in our history. At the time Europeans arrived in what is now America, there were some three to five *billion* passenger pigeons. It was, at the time, the most abundant bird on our planet. In his book *Hope Is a Thing with Feathers,* Christopher Cokinos asks us to imagine how large a single flock, observed by the naturalist Alexander Wilson in the early 1800s, would be. Wilson observed a flock of passenger pigeons a mile wide and 240 miles long. He estimated that there were no fewer than three pigeons for every cubic meter of sky, and this led Cokinos to figure the following: if every pigeon was 16 inches long, the 2.2 billion birds in this single flock would be equal to 35 billion inches, or about 3 billion feet. That's 563,200 miles worth of pigeons, enough to stretch around the equator 22.6 times. That, my friend, is a lot of pigeons.

The birds flew in flocks that sometimes covered the entire sky from horizon to horizon. They flew so closely together and in such large numbers that they literally blocked out the sun. They flew at the incredible speed of sixty miles an hour, yet despite their speed, a single flock might take three days to pass overhead.

The passenger pigeons fed on what biologist call "mast," the nuts produced by oak, chestnut, beech, and other trees. These trees produce nuts in different quantities at different times of the year, so the pigeons had to be nomads in the truest sense of the word. They were also opportunistic. They would feed on corn, berries, seeds, insects, worms, and snails. And they were voracious. Naturalists reported finding seventeen acorns in a single bird's crop. A crop could hold over a half-pint of nutritious beechnuts or one hundred kernels of corn.

While some pigeons did breed in small groups, most of the population gathered in huge nesting colonies. The average nesting colony covered about thirty square miles, but some of them spanned areas that are as hard to imagine as their total population size. The largest known nesting colony was observed in 1871 in Wisconsin. It covered 850 square miles.

The flock reproduced in what biologists term *synchrony.* That is, the huge flocks would descend as one, court, and build their nests over a three-day period, and then, on the same day, each female would lay one egg. Thirteen days later the eggs would hatch, and then the blind, naked hatchlings would spend two weeks in the nest. The parents took turns feeding the babies with "pigeon milk" until the young were old enough to feed on their own. Pigeon milk is something unique to the pigeon and dove family. It resembles white curd and flows out of the adult's crop when the young bird places its bill in the parent's open beak. After two weeks of tending the babies, called squabs, another spectacular event occurred. After gorging the squabs on one last meal, all of the adult birds rose as if on cue, formed their giant flock, and abandoned their young. The squabs, their crop as big as their bodies, flailed and fluttered, staggered on the ground, foraged for whatever they could find and, eventually, slimmed down and took to wing, forming their own giant flock. A roost was bedlam. The noise was staggering, and the stench, caused by piles of bird dung two to three feet thick beneath the trees, was over--

whelming. Yet in just over one month, the birds would all disappear in two great waves, first the parents and then the young, and there was no way of telling when or if they would be back again.

On the American frontier, passenger pigeons were a dietary staple and were hunted by American Indians and white settlers into the early 1800s without any noticeable decline in numbers. Most were hunted for the table, but commercial hunting was present from the earliest days. Records show, for example, that on a single day in May 1771, the Boston market had fifty thousand pigeons for sale.

It was not until the mid-1800s, though, that commercial hunting really came to the fore. Two key technological developments made pigeon hunting viable on a commercial scale. First was the development of the rail system, which enabled hunters to get the birds to market quickly and in large numbers. Second was the invention of the telegraph, which enabled hunters to receive reports on the exact location of the huge nesting sites. When a flock was located, two different systems of hunting the birds could be employed—shooting them or netting them. To shoot the birds, hunters would first attract the flock to the ground by clearing an area and spreading grain or other seeds. They also used live birds to lure a flock in with "fliers" (live birds that flew out on the end of a long string) or "stool pigeons" (live birds that were tethered to a special perch called a stool—we now call anyone who betrays their brethren by the same term).

The numbers of birds killed for the market (or to use as live decoys for hunters to practice trapshooting) were staggering. For example, most of my family now lives near Plattsburgh, New York. They would probably be surprised to learn that the 1851 pigeon harvest from Plattsburgh alone was 1.8 million pigeons. The Grand Rapids, Michigan, nesting of 1860 yielded some one million birds, and a pigeon hunt in Monroe County, Wisconsin, shipped two million birds to market in 1883.

But, by the 1880s, it was clear that the party was coming to an end. In fact, by 1886 there were only two giant flocks left in America, one in Oklahoma and one in Pennsylvania. The last known nesting in the northeast took place in 1880, in the mid-Atlantic states in 1889, in the Midwest in 1893, and in the Great Lakes region in 1894. Pigeons were still found, but not in the giant flocks that characterized their natural state. The last legitimate sighting of a wild passenger pigeon was in Ohio in 1900. (The bird was shot and mounted. If you want to see it, you can visit the Ohio State Museum.) By 1909, the last three passenger pigeons on the planet, two males and a female, lived in the

Cincinnati Zoo. The two males died, and by 1910 only the female was left. Affectionately named Martha (after the wife of George Washington), she lived four more years. Martha died on September 1, 1914, at one o'clock in the afternoon. This is one of only two instances in history where we know more or less the exact moment of a species' extinction. The only other instance, at least that I know of, was when the last member of one of the six species of *Partula* snails from the island of Moorea died at the Zoological Society of London, while under the watchful eye of Paul Pierce-Kelly. Paul has acted as the global coordinator of the *Partula* breeding program since it was first conceived in early November 1987 (several months after Ulie Seal delivered his lecture on the plight of those snails here at the Saint Louis Zoo). Over the years, the populations of the six species of *Partula* housed in zoos have periodically risen and crashed, but five of the species have done, on balance, very well. One species, *P. aurantia*, was limited to the London Zoo, and the population was stricken with a microsporidian parasite. Paul came to the zoo on a cold day, January, 1, 1996. One animal was left alive. It died later that morning.

In any case, we humans have always been extraordinary hunters. Even with the simple stone tool technology of the Paleolithic, when humans migrated into new areas, they had a dramatic impact upon animal life. Many people don't realize that the first massive wave of extinctions in North America occurred at about the same time that humans crossed the Bering Strait and began to migrate slowly down the Pacific Coast and then spread throughout the New World. At the time, North America was home to mammoths, mastodon, saber-tooth tigers, a giant sloth, enormous cave bears, a beaver species that was huge, and vast herds of horses. Shortly after the arrival of humans, all of them disappeared.

It was not just in North America. South America had similar mass extinctions about eleven thousand years ago, as did Australia about forty thousand years ago. In the case of Australia, not a single species larger than a human survived. Their giant sloth, rhinoceroses, lions, an automobile-sized tortoise, and the giant emu all disappeared shortly after the arrival of humans on the continent. In every case, it is hard not to assume that humans played a key role in the demise of these giants. As we've seen, with the development of agriculture in the Neolithic, our ability to cause massive environmental change only intensified.

Most, perhaps all, anthropologists acknowledge that the level of technology that a group has goes a long way toward determining the lifeways of a people. For example, the simple Paleolithic tools of the early hunters and gatherers in North America and Australia allowed these peoples to hunt very large animals successfully. These humans probably lived in small bands of closely related individuals. Group size was small, and the little bands were transitory, constantly roaming the countryside in search of food. A headman probably led those bands, but decisions were made by the consensus of the entire group. They probably had a strong ethic of sharing and a very limited idea of private property. There was undoubtedly no notion of ownership of land or water. Religious beliefs were probably what we call *animistic* in nature—that is, they probably attributed spiritual qualities to many of the animate and inanimate things around them.

As technology changed, first with the invention of agriculture and then with the Industrial Revolution, our ability to transform the environment changed with it. If we could graph technological change or development on this piece of paper you're reading, we could imagine the axis running across the bottom of the page as representing time. Let's start at about three million years ago on the bottom left side of the page and have the present on the bottom right. For the first 2,986,000 years, the line representing the rate of technological change was basically flat. It hardly rises from the bottom of the page and travels almost all the way across. Then, a mere fourteen thousand years ago, almost when our imperceptibly rising line hits the end of the page, we get the development of some new types of stone tools, characteristic of the Mesolithic period. Then, only eleven thousand years ago, there occurs an explosion in technology. With the Neolithic Revolution, we begin to develop an enormous number of new technological innovations in every field imaginable—metallurgy, transportation, weaponry, architecture, mining, navigation, communication, manufacturing—the list is almost endless. Now technological change is no longer *arithmetic* in nature. Suddenly it is *exponential* in nature. Instead of a situation wherein each development helps pave the way for one more, as happens for most of the first three million years of human development, we suddenly have a situation where each innovation paves the way for ten more, and each of those for ten more still, and each of those for even more. Now our imaginary line is shooting straight up. It traveled across the bottom of the page almost flat but, just as it almost reached the end of the page, it abruptly shot to the top.

Ah, but it didn't stop at the top of the page. It kept shooting upwards; who knows how far? Try this little experiment. Pick up a rock that has been chipped along part of its edge to make a rough point. That is a pretty good example of the level of human technological development over the vast majority of our history on this planet. Now, take this rock and compare it to your automobile. Obviously, the automobile is far more sophisticated, but is it a thousand times more sophisticated? A million times more sophisticated? Now consider how the automobile was made. Think of the tiny computers that control the engine, the coating that protects the metal, the plastics that make up many of the parts, the industry that must produce and refine the fuel to power the car, the great drilling rigs that must get the petroleum from the ground. The list seems to go on forever. Now think of the network of roads that span our country. Think of the power grids that produce the electricity to operate our traffic lights. Think of the colossal dams that must be engineered to harness the power of water to make electricity or the nuclear devices that might also be used for the same purpose.

I have no idea how far off the top of the page our imaginary line goes. Twenty feet? One hundred feet? Those are just a few of the questions we might ask if we stood next to our car. What if we stood next to a Boeing 707 or the lunar rover? What if we stood next to a surgical suite in a modern hospital? What if we stood next to a pharmaceutical plant or visited a genetic engineering lab?

What if we added all of those different places together? What if we took every single aspect of technological change and somehow figured out a way to measure it? How high would our line go? For the better part of three million years, technological change crept along with almost no significant new developments. Three million years are represented by the few inches along the bottom of this page. And now we have no idea how many feet in the air our imaginary line would go.

The thing is, even though our technology developed (and continues to develop) exponentially, our social and ideological systems still develop arithmetically. Think about it. Name me a single field of human social or ideological endeavor that has changed much at all over the last few thousand years of explosive technological change. Government? The ancient Greeks had tried just about every system that we currently employ today. Despotism, democracy, plutocracy, and dictatorships can all be found dating back thousands of years. Economics? There have been some interesting experiments in socialism, capitalism, and communism, but every single one of those economic systems can be found in the ethnographic record. In other words, all of them have likely been tried long before Eastern and Western cultures attempted them on a larger scale. Religion? Certainly the rise of monotheism and the ascendance of at least two great religious traditions in Christianity and Islam would be marked as major developments, but the ancient Egyptians had a prolonged experiment with monotheism long before worship of one God regained currency in the Middle East. Philosophy? Sure, we've added a number of chapters to the book begun by people like Plato, Socrates, and the brilliant philosophers of ancient China, but has there been change on a scale equal to that of the development of technology? Not hardly. Ethics? Most of the questions raised by Marcus Aurelius still confront us today.

The only area that I can think of that has changed as much as technology is technology's near relative, science. Science has been absolutely transformed. I could have a conversation with Socrates that would bring him up to speed on the major developments in philosophy over the last few thousand years and it wouldn't last long at all. On the other hand, trying to bring Aristotle, who wrote extensively in the areas of physics, astronomy, and the life sciences, up to speed would take years—if it could be done at all.

But that aside for a moment, our ability to understand and control the direction of our lives dropped abruptly some eleven thousand years ago. Our technological capabilities suddenly began to develop at a breathtaking pace, but our social forms—our family lives, political structures, economic systems, religious lives—are not much changed over the last few thousand years. Nor has our ideological system changed appreciably. If I could speak Greek, I could have a perfectly wonderful conversation with Socrates. He would probably know considerably more about what we have come to call philosophy than I would, and I certainly wouldn't be able to tell him anything about philosophy that he couldn't immediately comprehend and expand upon. Granted, our lives are much different now than they were for the little bands of hunters and gatherers that spread across the planet during the first three million years of human prehistory, but for the last few thousand years, our social and ideological lives have changed but little.

The question is, "If our social forms and ideological forms have changed very little, but our technology has changed enormously, what makes us think that we can comprehend and wisely direct technology?" The simple answer is that we can't. We can't because our strategy, whether we realize it or not, has been largely to use technology to try to solve the very problems that technology has created. Technology has deprived vast numbers of people of pure water, clean air, and beautiful vistas. It has transformed pristine landscapes forever. It has caused us to overconsume our natural resources, be they fish, forests, mineral deposits, fossil fuels, or the other bounties of our planet on a scale that is, simply put, unsustainable. The answer simply cannot be *more technology*.

But science is different. Scientific thought has progressed at much the same pace as technological change. Perhaps it was a little slower in the early going, but in the last few centuries it has come charging along. Now the pace of the development of scientific knowledge and awareness, if we could graph it, would be shooting up at the same rate as technological innovation. That is, I think, very good news for humanity.

But we must force ourselves to listen to what scientists have to say. Virtually all of them are saying the same thing about the greatest problem that confronts us as a species. They are saying that the transformation of natural landscapes must halt and that large tracts of habitat, particularly the richest and most diverse habitats on the planet, must be set aside and, ultimately, connected via wildlife corridors.

They are also saying that we must make changes in the way we decide to live our lives as individuals. The best way to describe it is that we must, as individuals and as a society, decide to reduce what scientists call our "ecological footprint." Simply put, our ecological footprint is the sum total of all the land needed to produce all of the resources required to sustain our lifestyle— food, energy, and materials—and to absorb all of the waste that we generate. It takes an *average* of 6.9 acres to support each of the world's people. But the average *American* requires 24 acres. We can still have a wonderful lifestyle and decrease our ecological footprint. The average German, for example, requires only 12 acres. On the other hand, the average Peruvian requires only 2. Americans must decide to tread more lightly upon the planet.

Let me give you some examples of what I mean. For hundreds of years, the Great Wall of China was the largest man-made structure in the world. In 1991, it became the second largest. The largest is now the Fresh Kills Landfill, which serves New York City. While we're picking on New York, it takes seventy-five thousand trees to produce one edition of the Sunday *New York Times*. If Americans recycled just one-tenth of our newspapers, we would save twenty-five million trees a year. We Americans now throw away 2.5 million plastic bottles an hour. Recycling alone would dramatically reduce our footprint. For every glass bottle recycled, we save enough energy to light a hundred-watt light bulb for four hours.

There is some good news. American steel recycling saves enough energy to heat and light eighteen million homes. Aluminum recycling is up, too, and it takes 95 percent less energy to make aluminum through recycling than it does by producing it from its natural ore, bauxite.

Still, our lifestyle causes us to take more than our fair share, and in the end, it is the biosphere that will suffer. In the end, we simply cannot rely on technological advances to reduce our footprint. Only by changing our society and our ideology, our values, beliefs, and attitudes, can we reduce our impact upon a besieged Earth.

Technology is a marvelous thing. We certainly use every technological weapon in our arsenal to help us do our work in zoos. We use satellites, telemetric eggs, and radio collars to track and document the habits of the animals that we care for. We use medical technologies undreamed of fifty years ago, and our ability to analyze nutritional requirements—rudimentary at best fifty years ago—is completely different now. Computers can help us to manage the genetics of small populations, a concept that was completely alien not long ago. We can analyze the complex endocrine systems of animals ranging from elephants to naked mole rats . . . but technology in and of itself will not save a single species from extinction. Indeed, if unabated, our technology will be our demise.

When I worked at the St. Louis Science Center, I got a major grant to restore a marvelous piece of technology that was left standing at the McDonnell Planetarium after the Museum of Science and Natural History purchased the building as the first phase of its transformation into the St. Louis Science Center. The piece of technology was a Thor-Able missile, sometimes referred to as the "workhorse of the space age." This particular type of missile launched some of the most important satellites ever placed in space—satellites that revolutionized telecommunications, our ability to forecast the weather, and our ability to monitor the very condition of our Earth. Such missiles also could be (and were) fitted with nuclear devices—devices that could, if deployed, destroy our planet. My job was to write a description of this missile for our visitors to read, and the text I wrote pretty much said what I just said. It concluded by saying that this missile is a perfect symbol of technology, in that it could be employed for great good or for great evil. The choice, of course, is always our own.

The trustees of the Science Center didn't like it one bit. They said that my job wasn't to scare people to death. But they were wrong. It is.

6

A Virus among Us

Most people are surprised to learn that over 95 percent of the animals in any given accredited zoo in America have been bred in zoos, many of them in highly selective, computer-driven breeding programs call Species Survival Plans (or SSPs). Now, that sentence requires more than a little explanation. First, it is important to understand the difference between an accredited zoo and the many other kinds of organizations that keep and often breed exotic animals. Second, it is important to understand just exactly what an SSP is and what it does. I have to admit that both topics—accreditation and SSPs—are not, in and of themselves, very exciting. But there is a very important relationship between Species Survival Plans in accredited zoos and the survival of *our* species, human beings. So here's the deal. If you'll read just a few pages about accreditation and SSPs, I'll tell you a wonderful detective story about how a curious vet at a famous zoo discovered a lethal virus—a virus that does not discriminate between many of the animals that live in our zoos and you. In 2002 this virus killed 284 humans, 4,300 horses, and hundreds of thousands of birds. That's a lot of deaths, even though not too many of them were humans. But it is only one virus, and many more will come. All of them will affect different species in different ways, and many of them will be a threat to you. This particular virus, if you haven't yet guessed its name, is called West Nile.

So, first things first. What exactly is an accredited zoo? Accredited zoos are zoos and aquariums that are members of the AZA (the American Zoo and Aquarium Association). The AZA was founded in 1924 and is a nonprofit organization dedicated to the advancement of zoos and aquariums in the areas of conservation, education, and science. Beginning in the mid-1960s, legislation was passed in several states that created a system of inspection and licensing of zoos and aquariums. On the federal level, the passage of the Endangered Species Act, the Animal Welfare Act, and the Marine Mammal Protection Act also reflected the public's growing concern for animal care. By 1965, the zoos of Great Britain had created a certification program that was designed to ensure that zoos in their newly formed zoo federation were adhering to the highest standards of animal care and ethics. In 1971 the AZA began the process of creating a similar program for North America, mostly in response to public concern over the future of endangered species housed in zoos. After three years of intense work, a new self-evaluation program had been developed and put in place. For eleven years the program was strictly voluntary but, by 1985, it became mandatory for all member zoos. In the first round of accreditation, the AZA lost about one-third of its membership, but nearly all of those evaluated their shortcomings, changed their programs, were reevaluated, and were readmitted. As I write this, there are 214 accredited, AZA zoos in North America.

Accreditation begins with a six-month period of self-study for the zoo, followed by a two- or three-day inspection by professionals from another zoo. Once those two things are completed, a special commission reviews all the supporting documentation (which can fill several large crates) and decides if a zoo or aquarium can be accredited. If so, the institution can go for five years until the whole process is repeated. I recently had a colonoscopy, a fiftieth birthday present from my wife. I would describe accreditation as being just about as invasive. It's not as bad as, say, the Spanish Inquisition, but at times it comes close. Still, it is vital to ensuring that zoos and aquariums adhere to the highest possible professional standards and, more importantly, since the requirements get tougher every year, it ensures that we keep doing a progressively better job.

The manual on accreditation approaches the size of this book, but it can be boiled down to eleven different areas of inquiry. Perhaps the most important deals with animal care. In order to be accredited, zoos must have a plan for how they will manage their collection, they must keep excellent records on daily care, they must adhere to approved policies for acquiring new animals or transferring animals out of the zoo and must keep records of those movements, they must display animals in approved exhibits, they must house them in approved off-exhibit areas, and they must provide enrichment for the animals in the zoo, making their lives as full and rewarding as possible.

Second, they must provide for first-rate veterinary care. Coverage must be available twenty-four hours a day, policies for drug use must be in place, veterinary plans for handling animal escapes and other emergencies must be in effect, animal diets must be tracked and monitored, and there must be a system for disease control in place along with a proactive system of preventative medicine.

Third, zoos and aquariums must be involved in conservation. This means that conservation must be integral to the mission statement of the zoo. Plus, if there is a Species Survival Plan for an animal in the collection, the zoo must participate in that plan. Finally, the process requires that the institution be "green" in terms of its daily operations. That means, to the maximum extent possible, we need to "walk the walk" in terms of recycling, water conservation, energy consumption, environmentally friendly construction techniques, and so on.

Fourth, zoos must have established education programs. They must have policies for using live animals in those programs, and the programs have to be evaluated for effectiveness.

Fifth, zoos must have written policies that document how animals may be used for research. Those policies must cover acceptable methods, staff involvement, evaluation, and guidelines for publication of results.

Sixth, there is a detailed inspection of the zoo's physical facilities. The accreditation team looks at general cleanliness in the public and animal areas and inspects the quarantine area. (All animals that come into the zoo are held in quarantine for at least thirty days to ensure that they are healthy. The word *quarantine* comes from the Italian word for "forty"—during the time of the plague, ships coming into Italian harbors were required to remain at anchor for forty days before any crew were allowed on shore, to ensure that no one on board was sick.) Food preparation areas are inspected, as are life-support areas, alarms and security systems, and of course, the exhibits and holding areas are all reviewed.

Seventh, safety and security programs are reviewed in detail. In general, zoos must have plans for dealing with animal escapes, responding to health emergencies for our staff and visitors, interacting with local police and fire departments, and so on. Zoos that display venomous or very dangerous animals such as large carnivores, must have even more detailed plans. Good communications must be in place, and security must be present on the grounds twenty-four hours a day, seven days a week.

Eighth, the accreditation team must review the entire staff in terms of their training, qualifications, and responsibilities. They also review staff size and compensation. (I've never heard of the commission saying that anyone was paid too much, so I really can't complain about their interest in ensuring appropriate compensation and benefits. This really is a profession that no one goes into for the money!)

They also look at governance. They want to know if our boards are looking at the right things in the right way. This is actually a fairly enlightened thing to do. We used to look at the function of boards in terms of what we called the "three Gs"—*give* us money, *get* us money, or *get* the heck out of here. Nowadays it's more like the "three Ws." We really want boards to share their *wisdom, work* hard on behalf of the institution, *and* to share their *wealth.*

Tenth, the accreditation team looks at the institution's support organizations. In particular, they want to know if the support organizations, like St. Louis's Friends of the Zoo, share our mission and goals and work constructively to benefit the zoo.

Finally, they take a very careful look at the zoo's finances. Are we financially stable? Are we providing for the long-term future of the organization? And, of course, they investigate to make sure that we are responsible in terms of our accounting practices.

I've been inspected three times over the last twelve years, twice in Indianapolis and once here in St. Louis, and I have also been involved in the other side of the process, as an inspector, several times. It's hard work for the institution and for the team that has to come in, work like crazy for two or three days, and write a lengthy report after spending several days poring over the zoo's required submissions. But it is still the best thing we have going.

The shame is that there are so many roadside attractions, so-called sanctuaries, and a host of other places that house exotic animals and don't come anywhere close to meeting the basic requirements for an accredited zoo. In fact, probably less than a tenth of the so-called zoos and sanctuaries licensed by the US Department of Agriculture (USDA) are actually accredited. While many of these facilities do valuable work and care adequately for their animals, many don't, and they should be closed down. In order to display or house exotic animals, all facilities, including zoos, are inspected by the USDA. Their inspections are nowhere near as detailed as the AZA's and, even worse, there are just not enough federal inspectors to really check on every facility as often as they should. Fortunately their abuses often come to light. Unfortunately, accredited zoos are often tarred with the same brush.

I mentioned earlier that several pieces of federal legislation marked the beginning of the public's interest in ensuring proper care for wild animals and, in particular, for endangered species. Perhaps no piece of federal legislation was more important in this regard than the passage of the Endangered Species Act in 1973. This act was designed to help species that were coming perilously close to extinction both in the United States and around the globe.

There are certainly problems associated with this law, but I would argue that they are less a function of the law and more a function of how the government has decided to implement the law. The law is pretty simple, but the federal regulatory bodies charged with implementing the law have added layer upon layer of complexity.

Here's an example. There is a beautiful blue butterfly, called the Karner blue, which was once found throughout Indiana and in neighboring states in a great arc from Maine to Minnesota. It is more or less gone from Indiana now (and just about everywhere else, for that matter, although it is still holding on in Michigan), but there is a small population breeding in a trailer park near Gary, Indiana. The Karner blue, like other species of butterflies, needs a specific host plant to complete its life cycle. In this case, the butterfly requires the presence of a beautiful perennial called the wild blue lupine. While I was at the Indianapolis Zoo, we approached the US Fish and Wildlife Service (USFWS) about reintroducing the Karner blue in the state. Our plan was simple. We would breed the butterfly, find landowners who would agree to plant the lupine, and then release the butterflies back throughout the state. "Fine," said the USFWS, "but remember this: if anyone agrees to plant the lupine and have these pretty little butterflies on their land, they also have to agree never to cut down their lupine. Ever. If they do, we'll describe it as a 'take' or an illegal killing of an endangered species, and we'll prosecute them." Now some landowners would probably be fine with that, but many wouldn't. Some would keep the lupine forever, but some might say, in effect, "What if my children's children want to plant a Christmas tree farm on our land?" The net result is that, instead of having thousands of Karner blues gradually repopulating the state, we have none. All because, for regulatory agencies, an absolute rule is easier to enforce than a relative one. As we'll see later in this book, the USFWS eventually came to terms with the zoos and conservation organizations that wanted to work with this gorgeous little guy, and there is now a restoration plan that two Indiana zoos, along with partners from many other states in the butterfly's range, are implementing. Still, in the initial stages, the government was far more of a hindrance than a help.

The Endangered Species Act also had a profound effect on zoos. It meant that we had to get specific permission from the federal government to bring in any endangered species to the United States. From 1973, we had to make a concerted effort to make breeding programs largely self-sufficient. Of course, early on in their history, zoos were consumptive users of wildlife. If a zoo wanted an animal, it would simply go out and get it. If it died, a zoo could always go out and get another one. While it was clear from the beginning of the century that many animals were going extinct, the extent of species loss was still not fully appreciated, nor was the role of humans in causing extinction fully acknowledged. As we have seen, the demise of the passenger pigeon is a classic case in point, and there are numerous other examples. The last well-documented sightings of the ivory-billed woodpecker were in a tract of land owned by the Singer Sewing Machine Company in the late 1930s and early 1940s. That tract was logged shortly thereafter, destroying any hope for the continuation of the species. (Rumors of ivory-bills continued to persist; in 1987, Jerome Jackson, one of the world's authorities on the ivory-bill, heard a pair calling in a forest north of Vicksburg, Mississippi. That forest was later cut down, too. As this book goes to press, conservationists are reeling with the amazing discovery that at least one ivory-bill indeed perseveres in the Big Woods of eastern Arkansas.)

The point is that, until fairly recently, no one cared, and to this day, many people simply don't care enough. Respect for our human obligation to preserve wildlife, together with the awareness that many species could be destroyed forever is, in the grand sweep of history, a very new idea. By the mid-1970s, the role of zoos in preserving wildlife through breeding had come to be fully recognized, but zoos had to change along with the rest of society. The Endangered Species Act codified that need and the role that zoos would need to play if we were ever to have a hope of reversing the decline of species worldwide.

There was, however, a problem with breeding populations in zoos. This problem was described in a 1979 publication by Katherine Ralls, Kristin Brugger, and Jonathan Ballou of the National Zoo. This publication documented the dangers of inbreeding by showing that it was highly correlated with juvenile mortality in zoo collections. Their original research was based on sixteen species of hoofed animals; they found that in fifteen of the sixteen, the babies were more likely to live beyond six months if the parents were unrelated. A later study, which covered forty-four different mammals, confirmed the earlier conclusions and some of the species studied showed striking problems. For example, for the scimitar-horned oryx, babies that weren't inbred had a 5.4 percent mortality rate. For inbred oryxes, 100 percent died within the first year. By the way, the scimitar-horned oryx eventually went extinct in the wild. Since then, it has been successfully reintroduced to the wild by a collaborative of zoos and other organizations and is now holding its own.

Many people would be surprised that this conclusion wasn't documented until 1979, but again, it is easy to forget how quickly scientific discoveries become integrated into our everyday thinking. When I was studying genetics as a young anthropology student, one of the great debates in the literature centered on the genetic aspects of incest, or inbreeding in humans. There was one school of thought that maintained that, in the long run, inbreeding could be beneficial because it would result in the loss of individuals that carried deleterious recessive genes. The argument ran something like this: individuals may suffer, but the population on the whole might be better off. After all, Cleopatra was the product of eleven generations of incest among the Ptolemaic royalty, and she, rumor has it, was a pretty good looking young lady. The simple truth is, we didn't know for sure how detrimental inbreeding was for smaller populations until it was studied in zoos. I once asked Dr. Ballou (in the best tradition of the Watergate Hearings), "What did you know and when did you know it?" In other words, I wanted to know if he suspected that inbreeding was a problem all along and was just doing this study in order to confirm his suspicions. His answer surprised me. "Well," he said, "we thought we could find the exact opposite. For years, all of our suppositions were based on what we knew from domestic breeding, where animals are closely bred for positive traits. On that basis, you could assume that inbreeding could be good. On the other hand, we were seeing very high infant mortality." In the end, the evidence was clear. Inbreeding was highly correlated with premature deaths.

Once we did know for sure that inbreeding had a horrible impact on infant survivability, we began to coordinate breeding in a systematic fashion through Species Survival Plans. These started in 1981 as a cooperative population management program among all accredited zoos. The idea was to maintain a captive population that was both genetically diverse and demographically stable. By 1989, there were 50 species in the program, and now

there are 106 SSPs covering 161 individual species, 10 of which are administered at the Saint Louis Zoo. (There are more species than there are SSPs because many SSPs, like the one for *Partula* snails, actually have three or four
very closely related species in a single plan.) The programs cover charismatic
species, like the cheetah, an animal that is literally racing to extinction, as
well as less alluring animals, like the Puerto Rican crested toad, which, like
many amphibians worldwide, is in dramatic decline. The Saint Louis Zoo
just released eight thousand crested toad tadpoles back in the wild after years
of coordinated breeding. Our biologists were successful in breeding crested
toads only after they figured out how to simulate the dry spells, followed by
torrential rains, which characterize the climate of Puerto Rico.

A species must meet a number of criteria to become part of this program.
For the most part, they must be endangered species that are interesting to the
qualified zoo professionals that will have to figure out how to breed, rear,
and, in many cases, release them back to the wild. Usually they are what we
call "flagship species," like the giant panda, California condor, and the lowland gorilla. Flagship species are "popular" enough that people will support
our efforts to save them, the theory being that if we can save the species *and
the habitat it lives in,* we can save all of the other (perhaps less popular) species
from the same area.

Each SSP has a volunteer coordinator at a zoo who is responsible for day-
to-day activities. The coordinator, along with a committee composed of representatives from other zoos, is responsible for population management,
research, education, and in some cases, reintroduction of the species back to
the wild. All of the SSPs have a master plan that is based on a family tree,
which helps the coordinator figure out which animals should be bred together. The goal is to maintain at least 90 percent of the existing genetic vari-
ability for one hundred years out in the future. At the same time, those plans
also have to determine which animals *shouldn't* be bred together so that the
population won't outgrow the available space in the nation's zoos. The coordinator walks a fine line. With charismatic species like gorillas, a recommendation that a very popular animal be moved from one zoo to another can
cause a surprising amount of public dismay. The coordinator might also recommend that an animal not be bred. In the not too distant past, zoo directors
universally hated that. Babies made for good press coverage every spring,
and it was hard to let go of the PR value that they brought.

SSPs are, in turn, organized into Taxon Advisory Groups (or TAGs). With
TAGs, experts from zoos are asked to make recommendations about breed-

ing and conservation regarding similar *groups* of animals, like marine fishes or penguins. So, while there is an SSP for the Humboldt penguin, there is a TAG for penguins in general. At the risk of dipping us into the hot broth of conservation alphabet soup, these efforts are integrated by geographical region into CAPs (Conservation Action Plans).

One more acronym, and then I'll stop. SSPs exist in North America, but other regions of the world have more or less identical programs. (They all have different names, hence different acronyms, but I promised to only give you one more acronym, so I won't list all the names.) So, here's the last acronym: ISIS. It stands for International Species Information System, and ISIS is the group that ultimately links all of the world's zoos and aquariums, along with other organizations that care for exotic, threatened, or endangered species, into one big group. ISIS was formed back in 1974 with fifty-five members and has grown steadily ever since. Now it has more than six hundred institutions, in seventy countries, covering six continents. ISIS tracks all the animals held in those institutions. That means ISIS has data on over 1.65 million animals, representing over 10,000 species. As I write this, ISIS is developing its fourth generation of software at a cost of about $10 million. The new software will be Web-based and will function in what the computer geeks refer to as real time. From a zoo's point of view, what it will allow us to do is to organize SSPs on a global basis. In other words, instead of having North American, European, and Australasian programs acting independently, we'll be able to act globally, vastly improving our ability, from a genetic point of view, to preserve all of these different species. In fact, it might be fair to characterize ISIS's credo not as "think globally, act locally," but as "think globally, act globally (but don't forget that local zoos are where it happens)." Not a very catchy slogan, but it gets at what we're trying to do.

The other thing that ISIS does is track all of the vet records for individual animals. Here's where zoo animals have it all over humans. I promise you that your medical records will never all be in the same place. In fact, most people in America couldn't find all of their medical records if they tried. For example, a few weeks ago, I caught a bad bug in Tanzania and a wonderful young Tanzanian physician gave me a prescription for something to stop my stomach from cramping. I don't even remember his name. I could never remember all of the physicians I've ever seen in all of the places I've ever lived or visited. But that's not the case for zoo animals. I can go over to the vet hospital, select a specific gorilla, push a button, and get a printout of every single procedure, exam, or test result for that individual gorilla over the entire course of the gorilla's life. In that respect, we can care better for a gorilla in a zoo than we can, say, for me. The records go from birth to death and follow the animal wherever it goes, even if it is sent, for example, to another zoo for breeding.

"Okay," you might say, "that's interesting, but why should I care?" The answer is, you should care because, if ISIS brings its new Web-based, real-time medical record system online soon, it could save your life. That should at least get your attention, but in order to explain how ISIS could save your life, I need to talk a little bit about a special class of diseases called *zoonotic* diseases.

Zoonotic diseases are ones that can cross from one species to another. More than half of the human pathogens fall into this category, and as we learn more, that percentage is going up. A good example is the flu. A flu virus can jump from an animal, for instance a human, to another animal like a pig or a duck, and then back to a human. While living in its new host—let's say it's in a pig—it can mutate. So your immune system, which was all geared up to fight the old virus, is suddenly confronted with a brand new virus that you don't have any immunity to. The reason we have so many strains of flu (and the reason they're often referred to as "Asian" flus) is that in many places in Asia (for example China), people live in very close proximity to animals like pigs and ducks. That makes it easy for the virus to hop and mutate. Plus, there are a lot of hosts (that is, people) living close together, so the virus has a nice big human reservoir and can spread quickly and widely. With modern air travel, the virus can quickly move from China to your son's day-care center, and then on to you.

There are many zoonotic diseases. It would appear, for example, that AIDS is one of them. Another one that you've certainly heard of is West Nile virus, and the story of how that virus was detected here in North America illustrates how accredited zoos, because of the wide range of species they house and the incredible level of medical care those species receive, can help safeguard you and your family.

West Nile virus was first isolated in the West Nile district of Uganda back in 1937. It wasn't until 1957, during an outbreak of the virus in Israel, that physicians fully realized that this virus could cause severe meningitis or encephalitis (inflammation of the spinal cord *and* brain) in elderly human patients. It was found in equines (members of the horse family) in Egypt and France in the early 1960s. In 1999, the same strain of the virus that hit Israel over forty years ago jumped to the United States.

About 80 percent of people who are infected with West Nile show no symptoms at all. The other 20 percent display symptoms that can include fever, headache and body aches, nausea, and vomiting. Those symptoms can last anywhere from a few days to several weeks. About 1 in every 150 people who have West Nile becomes very sick. People in this category would show many of the same mild symptoms, but their fevers can become quite high, they can become comatose, or they can suffer from tremors, convulsions, numbness, and paralysis. Even if they recover, some of those neurological effects may be permanent.

As is the case with many other zoonotic diseases, mosquitoes play a pivotal role in the spread of this illness. They become carriers when they bite another infected animal, in this case, a bird. Virologists spent a good deal of time studying West Nile as it made its way across the Old World, moving from Africa to Europe, the Middle East, and west and central Asia. They came to suspect that migratory birds were the main carriers because wherever there was a new outbreak in a temperate region of the Old World, the outbreak seemed to coincide with the arrival of large numbers of migratory birds in the late summer or early fall. Plus, the outbreaks seemed to occur among humans living in or near wetland areas where there were both high concentrations of birds *and* high concentrations of mosquitoes. The catch is, in the Old World, the birds and humans weren't dying at the same time. As we'll see in a second, that was one of the things that confused scientists when West Nile made its leap to our shores.

The presence of West Nile here in America was discovered by Dr. Tracey McNamara, then a pathologist at the Bronx Zoo. As we learned in the story about the fire in Philadelphia, every animal that dies in an accredited zoo is necropsied. The amount of time and energy that goes into a necropsy varies from species to species. For example, when an elephant dies in a zoo there are an enormous number of things we want to know. The same thing would be true of almost any primate. We also collect tissue samples from the dead animals and store them. We're not always sure why. It could be that there was something, for example, in the blood of a dead animal that we didn't know we should be looking for until years later. By collecting tissue samples, we can quite literally go back in time to investigate something like the spread of a disease, so saving tissues certainly gives the benefit of hindsight.

But it's not just our zoo animals that we necropsy. We look at every animal that dies in the zoo, or at least we try to, even if that animal crawled, slithered, or flew in on its own. The reason is obvious—it could be carrying a disease that could spread to other animals in the zoo.

In August 1999, Dr. McNamara began to get dead crows that staff members had found on the grounds of the zoo. She knew that single-species die-offs in birds are rare, so she was immediately curious. By August 25 she was regularly shipping the dead birds off to New York's Department of Environmental Conservation, but she wasn't getting any response, so she began to do her own dissections. What she found was that all of the crows were bleeding from the brain. She first suspected an outbreak of one of the two diseases that can cause massive die-offs of birds—Newcastle disease or avian flu. But, she reasoned, if it was either of those two diseases, the zoo's chickens and turkeys would be dying also, because they're both highly susceptible. Since they were all healthy, she moved on to Eastern Equine encephalitis. Not a bad guess, but it turns out that emus, an Australian bird that the Bronx Zoo had in its collection, are highly susceptible to this disease, but again, all the emus were healthy.

Fast-forward to September 8. Dr. McNamara comes back from her Labor Day weekend to discover that, in addition to the crows that had flown in to the zoo and died back in August, now animals from her own collection are dying. In particular, she noticed that three of the four types of birds were New World species: a Guanay cormorant, three Chilean flamingos, and a bald eagle. Her dissections revealed the same brain lesions that she had seen the previous month in the dead crows.

She now knew several things. First, she knew, from the cultures she had tried to grow from the dead crows since the beginning of August, that the disease wasn't bacterial. Second, she knew that there was no evidence of pesticide exposure. Third, she knew that all of the dead birds were species that the zoo kept outdoors during the summer months. Fourth, she knew that the disease didn't look like anything she had ever seen in birds before. Fifth, and most important, since it was primarily North and South American birds that were dying, it looked like she might be confronted with something that was new to our hemisphere.

She also knew something else. She knew that people were dying in New York from what was thought to be a mosquito-borne disease. So far, we've just heard the story from the zoo's point of view, but Dr. McNamara has been reading the local papers, and what she's reading disturbs her. It simply doesn't make sense.

As early as June 1999, people were reporting dead crows in Queens. In July they started to die in the Bronx (and, as we've seen, by early August Dr. McNamara was picking up dead crows in her zoo). By the third week of August, Dr. Deborah Asnis, who worked at the Flushing Hospital in Queens, reported that she had three elderly patients with neurological symptoms. She thought that was unusual, so she sent samples of their blood and spinal fluid to the Centers for Disease Control (CDC) in Atlanta.

The CDC thought it was unusual too, so it began interviewing the families of the patients, asking about where they had traveled, what they had eaten, or anything else that might link them. The only thing the CDC found that they had in common is that they had all been spending a lot of time outdoors in the evening. That wouldn't have meant much by itself, but in the meantime, another case of what looked like encephalitis came to light.

When the CDC sees encephalitis-like symptoms during the mosquito season among people who have spent time outdoors in the early evening (when mosquitoes are most active), they assume that they're dealing with one of the North American mosquito-borne encephalitis viruses, like our own St. Louis encephalitis. St. Louis encephalitis made its first appearance in St. Louis in 1933 and has been present in North America ever since. There hadn't been an outbreak of St. Louis encephalitis in New York since the 1970s, but when the CDC tested the blood and spinal fluid of Dr. Asnis's patients for St. Louis encephalitis antibodies, they all tested positive. On September 3, the CDC announced that the culprit was St. Louis encephalitis. From the CDC's point of view, the case was closed.

But none of this made any sense to Dr. McNamara for the simple reason that St. Louis encephalitis doesn't cause die-offs of birds—and birds were certainly dying. She didn't care what the CDC said on September 3, she didn't care what the textbooks said, and for that matter, she didn't care that the tissue samples of the human patients were testing positive for St. Louis encephalitis. She knew that there simply had to be another explanation.

On September 9, two more flamingoes died at the Bronx Zoo. Now Dr. McNamara was really nervous. She feared that there was a link between the bird deaths and the cases of what the CDC thought was human encephalitis. She called the CDC and said that she doubted their diagnosis. Since St. Louis encephalitis doesn't normally kill birds, she thought that the CDC should take a look at the tissue samples from the zoo animals. The CDC refused.

Undeterred, she offered to send the same samples to the National Veterinary Services Lab in Ames, Iowa. This lab is run by the USDA, the same branch of government responsible for inspecting zoos. The CDC wasn't interested in her animal samples, but the USDA was. By September 13, the lab had confirmed what Dr. McNamara initially suspected. Their cell cultures showed that the same thing that was killing the wild crows was killing the zoo birds. On September 15, the National Veterinary Services Lab made a second key discovery. They examined the viral particles under an electron microscope and found them to be different from any known virus that had ever killed animals in the New World. The problem was that they couldn't do any more work with virus. They didn't have the proper containment facilities. Viruses are tricky things. Viruses that are lethal are particularly tricky. So the lab immediately called the CDC.

By September 19, the CDC still hadn't responded. At this point, I'm guessing that Dr. McNamara was feeling not only rejected but probably also downright mad. If the CDC wasn't interested, there had to be someone in government who was. So she called the US Army Medical Research Institute in Infectious Diseases, located in Fort Detrick, Maryland. It took two more days to get in touch with the right army scientists, but when she finally did, bells went off in Maryland that could be heard all the way to the CDC headquarters in Atlanta. The CDC changed its tune about Dr. McNamara's samples: "Maybe you should send some down."

But the US Army was already at work. By September 23, they had discovered that although Dr. McNamara's samples tested weakly positive for St. Louis encephalitis, they did not believe that this was the virus that was killing the birds. They asked for more tissue so that they could look for other diseases: Japanese encephalitis, Powassan, dengue fever, yellow fever, and West Nile. On September 24, they determined that it was West Nile. The CDC agreed. The conclusion was inescapable: it was not St. Louis encephalitis that was killing people in New York—it was an old disease that was new to North America.

In case you're wondering, we'll never know how West Nile made it across the ocean. There are several possibilities, but only two of them strike me as remotely likely. We don't think that the disease could have come from an infected individual that was then bitten by a mosquito, that, in turn, bit a bird—as humans, our viremia (the amount of the virus present in an infected individual) is not high enough to infect a mosquito. The disease could, however, have come across the ocean by ship—perhaps an old tire on the deck of a freighter had enough standing water that the mosquito larvae could survive the long voyage. Alternatively, the virus could have come in a bird that was illegally smuggled into the United States. The trade in exotic pets in general, and birds in particular, is having a devastating impact on wildlife worldwide. For example, many of those beautiful macaws that you see in pet shops probably got here illegally. There's tremendous profit for the smugglers, and it is fueled by public demand. If West Nile came from an illegally imported bird, then there is an element of poetic revenge in the West Nile story. Sometimes, at least, we humans are made to pay for our avarice.

So, how can ISIS help guard us against infectious diseases, and why did I make you wade through all that stuff about SSPs and accreditation? The answer to the first part of the question is that, currently, ISIS is not a Web-based, real-time system. If the medical records for all of the animals in the six-hundred-member global ISIS database were entered into one giant system, and they came into the system every day, then we would have a database that could monitor disease outbreaks, such as West Nile, anywhere in the world.

Remember, a similar system will probably never develop for humans. In the first place, our records are not in one place for any of us individuals. More importantly, even if they were, there is no health organization in the world that collects all of the individual records into one giant database. But what will never happen for humans is already almost in place for animals in zoos and aquariums.

What makes accreditation so important is that it requires zoos, at least in the United States, to do several critical things. First, we have to look at all of our animals more or less every day and keep records on their health, feeding patterns, or any unusual behaviors. Second, we have to provide regular veterinary care and keep all the records of the tests, procedures, and exams on each individual animal. Third, accreditation requires zoos to have good collection plans, and we house a wide range of species (about ten thousand different species in total, counting insects and fish). This means that we have sentinel species for a wide variety of different zoonotic diseases in our collection. This is important because some zoonotic diseases, like West Nile, strike some species (like humans, horses, and crows) and pass others by. Fourth, accreditation requires us to document not only the health of an animal during its life, but also the causes of its death. If we didn't necropsy every animal that died in our care, we wouldn't learn how to care for them better.

In fact, accredited zoos are the only class of organizations in the United States that necropsies every animal that dies. No other organization can make this claim. As a consequence, no other organization could serve as a better sentinel for zoonotic disease.

As I write this, ISIS has raised five million dollars from the world's zoos and aquariums to create this new software system, and we continue to lobby the government for additional support to make this system a reality. The computer programmers estimate that it will take another five million dollars to install the software around the world. I predict it will be worth every penny. There will always be viruses at large in the world around us; I, for one, would like to hear about them early enough to actually stop them from reaching me.

This royal flycatcher was caught in a mist net as part of the Saint Louis Zoo's animal census in Nicaragua. SAINT LOUIS ZOO

Amali, the first African elephant conceived by assisted reproduction, and her mom. RICH CLARK

The Guam kingfisher—one of the rarest
and most beautiful birds in the world.
CHUCK DRESNER

The Jamaican ground iguana,
defending his territory.
RICH CLARK

Samburu game scouts from Westgate, a community near Kalama, Kenya, where the Saint Louis Zoo works with Grevy's zebras. JEFFREY BONNER

Martha Fischer and Belinda Low putting a radio collar on a Grevy's zebra.
JEFFREY BONNER

Partula snails on a leaf at the Monsanto Insectarium.
CHUCK DRESNER

A rhino from the Kruger shows his incredible horns. JEFFREY BONNER

Dr. Randy Junge from the Saint Louis Zoo in his portable field lab in Madagascar.

A *ring-tailed lemur, one of the many species of lemurs from Madagascar.*
CHUCK DRESNER

Cheetahs from the plains of Tanzania. JEFFREY BONNER

Humboldt penguins on the coast at Punta San Juan, Peru, where
several zoos run a cooperative field station. JEFFREY BONNER

A tribesman from New Guinea, where the Saint Louis Zoo is
part of a community education program. MARK WANNER

A bongo from the Mount Kenya Safari Club, where zoos are participating in a bongo reintroduction program. JEFFREY BONNER

Jeff Ettling, of the Saint Louis Zoo, radio tracking Armenian vipers. SAINT LOUIS ZOO

An Armenian viper,
one of the most beautiful
vipers in the world.
MICHAEL JACOB

One of the Saint Louis Zoo's beautiful American burying beetles.
MICHAEL JACOB

A Missouri hellbender, the highly endangered amphibian from Missouri's fast-flowing streams.
SAINT LOUIS ZOO

Lions and Tigers and Bears

In 1993, North American zoos decided to form the first SSP for an animal that wasn't threatened or endangered—the African lion. Most of the lions found in zoos up until then were of unknown genetic heritage. We had been breeding lions in zoos without much regard for ancestry—because they weren't exactly rare or difficult to reproduce, and because they had been bred in zoos for so long, there simply weren't good records. The SSP concluded that it would be easier to just let all of the existing zoo lions die out and gradually replace them with lions of known parentage. Ideally, those lions would all come from the wild.

I was not crazy about the idea. Here we were devoting a considerable amount of energy to a subspecies, the African lion, that wasn't in the least bit of trouble, when we had another subspecies, the Asian lion, which was on the edge of extinction. People don't often think about it, but lions once ranged up through Greece and Macedonia through what is now Turkey and from there all the way into India. Remember all those fables where Greek guys are getting thorns out of the lion's foot? Well, they didn't take place in sub-Saharan Africa. They took place in Greece, with Asian lions.

Of course, they're mostly gone, now. In 2002 the numbers of Asian lions were estimated at 280, with most of those animals living in the Gir Forest in India. This is an area of about eighteen hundred square miles in northern India. With the growing human population in this region, though, the lions'

115

effective habitat is now down to about eight hundred square miles. We don't know how many animals there are for sure, because the ones we think might be Asian lions have probably been crossbred enough with African lions that they are no longer, in effect, "purebreds." Asian lions don't look much different when compared with African lions, so I'm not sure it makes much difference. Then again, a regular horse probably wouldn't look that much different from a unicorn. I'd still like to see a unicorn, though. The most visible differences between Asian and African lions are that the Asian lions have a fold of skin running along the abdomen, and their manes look a little different in the males.

The crossbreeding of Asian and African lions goes all the way back to Roman times. They were moving lions around the empire and breeding them with one another with no regard at all for parentage. In order to save the Asian lion from extinction, we could have organized a push to fund the census work, the necessary genetic testing, and the importation and breeding of the remaining Asians. In fact, we could still probably save them from extinction. The prevailing feeling, though, was that it is too much work for too little return and that we should just concentrate on African lions. I still disagree with the decision.

In the same year that the AZA decided to form an SSP for lions, the Indianapolis Zoo's only male lion died. The zoo's first male, Kitaba, lived to the age of nineteen. Since their life expectancy in the wild is quite a bit less than fifteen years, he was getting on in age. After his death, the zoo acquired a new male named Lionel from the Potter Park Zoo. He had been raised by a truck driver who had kept him in his cab for company (and probably as a deterrent to theft). The truck driver fed him live chickens for amusement. When he got big enough that he was no longer cute, the truck driver dropped him off at Potter Park. He died seven years later, leaving the Indianapolis Zoo without a breeding male.

At the same time, the zoo had two females that were nearing twenty years of age. Shortly after the death of the big male, we lost a female due to complications related to old age. But in September, we learned of two lions that were available for export to the United States.

Their names were Mwangi, which in Swahili means "he will have many children," and Shamfa, Swahili for "sunshine." They had been kept illegally on an African farm. When wildlife authorities found them, they were confiscated and sent to the Pretoria Zoo. Pretoria didn't have much room but agreed to keep them until a permanent home could be found. With the loss of two of our three lions, Indianapolis was suddenly an ideal candidate.

Their trip to Indianapolis proved to be an adventure. In February 1994 the two lions were loaded into crates and taken to the airport. Their flight was scheduled to arrive at O'Hare Airport on the sixth, but we got a call from our transport team in Chicago telling us that the plane never arrived. After some frantic calls, we discovered that a bird had been sucked into one of the jet engines after takeoff and the plane had immediately turned around to make an emergency landing. Then, on landing, the plane lost its hydraulics and blew out sixteen tires. Needless to say, Mwangi and Shamfa were a little banged up.

We decided to have them go back to the Pretoria Zoo for observation, and a week later we tried it again. This time they arrived in Chicago safe and sound and enjoyed a short three-and-a-half-hour drive to Indianapolis. When the crates were unloaded, however, we discovered that they wouldn't fit through the door.

They weren't very big lions at the time, but all the same, no one wanted to try to herd them from the crate into the holding area. Fortunately, our maintenance staff showed up with a tape measure and a screwdriver. After they had taken off the holding-area doors, the crate just fit through the doorway. The crate was bolted to the holding pen, the guillotine-style door was raised, and the lions entered their new home.

By 1996 they were old enough to breed. When lions mate, they mate a lot. For several days they engaged in pretty much nonstop sex, copulating up to 150 times a day. On April 15 the results of that activity appeared in the form of two gorgeous lion cubs. If you stop to think about it, there are five things that have to happen for a successful breeding effort. You need to have a compatible pair of animals; they must successfully mate; you need an unremarkable pregnancy; a successful birth; and finally you must have a nurturing mother. We had all five in abundance.

By the first week of June, the cubs were old enough for their first exam. Shamfa was shifted to an adjacent cage, and the babies were weighed (both had grown to sixteen pounds by then), dewormed, and vaccinated in much the same way that kittens are. The staff took turns holding them and all had their pictures taken with the babies, but everyone worked fast. Shamfa did not like being separated from the cubs for long and was soon pacing and growling her displeasure.

Lion cubs are simply hysterical to watch. They act like kittens but look more like giant puppies, with giant, gangly paws, boundless energy, and a zest for exploration. Adult lions become their playground. They swipe at their mom's tail, stalk her, and pounce with a kind of innocent ferocity that will, within a few short months, become a swift and lethal attack. They look adorable, but zookeepers are trained not to try to tame them. We want lions that behave like lions, and that means rearing them in as natural way as possible. Still, it's hard to resist these delightful little balls of fur, with their blue eyes and limitless personality.

By June 15, they were ready to go out to their new exhibit. The two cubs were named Sekaye, which means "happy with laughter," and Dumisani, which means "herald of the future." The latter name was particularly apt. These were among the first lion cubs born in the history of the new SSP. We all hope that we will never need to reintroduce lions into Africa, but if we do, we will now be certain that the lions in North America are genetically the same as the lions currently found in the wild. It is sad to contemplate the possibility of a future so bleak, but it could certainly happen. If you don't believe me, just look at the Siberian tiger.

We do not know for sure how many Siberian tigers are left in the wild, but we estimate that the number, in 2002, was about three to four hundred. Siberian tigers are now called Amur tigers; they have been renamed for the Amur River, which runs through Siberia for some twenty-seven hundred miles. They once inhabited the snowy Russian forests from north of Mongolia, past North Korea, and all the way to the Sea of Japan.

Amur tigers are the largest of all the big cats and arguably the most beautiful. They look much like Bengal tigers, but they are taller and have more whitish coloring on their heads and underbelly. They are huge, weighing up to eight hundred pounds. When our Amur tiger at the Saint Louis Zoo stands erect against his enclosure, his head is higher than I can reach with my hand over my head. They are masterful hunters, known for their grace, cunning, and power.

In May 1993 the Indianapolis Zoo got a new tiger from Siberia. When she came to Indianapolis, she was only a few months old, and her name was Nadirzda, a Russian word meaning "hope." Nadirzda's mother, named Lena, lived in the Sikhote-Alin International Biosphere Reserve in far eastern Siberia. This reserve is home to a very large concentration of the remaining Amur tigers—probably some two hundred in number. Two American researchers, Howard Quigley and Maurice Hornocker, had teamed up with Russian biologists to study the tigers in the preserve, combining their skills in tranquilizing tigers and fitting them with radio telemetry collars with the Russians' skills in tracking tigers in the wild.

In June 1992 the research team tracked and shot Lena with a tranquilizer and then fitted her with a radio collar. For five months, they tracked her movements through the forest until, suddenly, the collar stopped moving entirely. Researchers rushed to the spot, only to discover that Lena had fallen victim to poachers. The collar had been slashed off her carcass and left in the snow.

But what the poachers didn't know was that Lena had cubs. There were four little ten-week-old babies hidden nearby. Two of them, already weak from lack of food, had died. The other two were picked up by the scientists and locked in their small cabin outside the town of Terney. These two cubs were far too young to survive in the wild, so the researchers began a series of frantic phone calls to American zoos, searching for an institution that could take the cubs. They also made a series of phone calls to Russian officials in an effort to line up the necessary permits required to get them into the United States.

Clear-cutting of timber and poaching are the tiger's two greatest enemies. With the collapse of the Soviet Union, government officials turned to the logging industry in an effort to generate hard currency. Enormous expanses of the taiga—the spruce, fir, and pine forests that make up the Siberian forest—were leased to foreign corporations. The corporations clear-cut the land, totally destroying the tiger's habitat. In 1992 they were cutting about ten million acres a year. They also build roads to get the logs out, and these roads become the poachers' best friend.

The roads allow for easy access from China and North Korea into the tiger's range and give the poachers access to those lucrative markets. They sell every part of the tiger, including the skin, internal organs, and bones. The genitals, in particular, are thought to have great medicinal value. The bones are crushed and used to make "tiger wine." The total value of a poached tiger was, at that time, about ten thousand dollars, roughly equal to one year's wages.

While the cubs were temporarily safe, locked in the cabin in the town outside the preserve, it wasn't easy to raise the babies. Eventually the scientists found a supplier who had whole sides of beef for sale. These were cut up with an ax and injected with vitamins and minerals. Fortunately, the project had a vet on the team, and she proved to be up to the task of caring for the precious infants.

By January 10 of the following year, the permits were in place and the babies were flown to the United States. Both Nadirzda and her brother went initially to the Henry Doorly Zoo in Omaha, and then Nadirzda was flown to Indianapolis, her new home. In Indianapolis, her name was changed to Lena, in honor of her mother.

Lena hated me. I'm not suggesting that it was anything personal; she just hated my guts. She wasn't particularly affectionate with anyone, but she went quite a way in the other direction with me. Perhaps I resembled one of the poachers with my distinctive moustache and glasses, or perhaps she didn't like men who wore business suits. Perhaps I reminded her of a vet she had encountered along the way and she associated me with the discomfort of taking blood samples or having a physical exam. We'll never know for sure what caused it.

On the other hand, it was a good thing that she felt the way she did. Her adult canines were growing up over her baby canines, which hadn't loosened at all. Although the vets weren't too concerned, they did want to monitor the situation. That's where I came in. Whenever they wanted to do a physical, they'd call me into the holding area. As soon as Lena saw me, she'd open her mouth and hiss. The vets could peer in her mouth all they wanted, but Lena would never take her eyes off of me. It was gratifying to be so useful, even if I did feel like the proverbial turd in the punch bowl.

As a wild-caught tiger, Lena's genes were unrepresented in the North American population. The computer determined that, as soon as she was old enough, she should be bred to a tiger named Dshingis, who was then living at the Minnesota Zoo. Dshingis came to Indianapolis in 1996 in anticipation of breeding him with Lena as soon as she was old enough.

The tiger exhibit at Indianapolis was actually designed as a breeding facility. The exhibit consists of two viewing yards, so that other animals can be separated from the mother and cubs. One of the yards was specifically designed for cubs, with viewing areas that allowed the mother to hide from visitors if she was feeling threatened or insecure. The pool in the exhibit was designed so that keepers could adjust the water level until the cubs learned to swim, and the doors to the exhibit were designed so that cubs could get on and off exhibit easily.

Dshingis and Lena got along well, and June 1998 saw Lena at the height of her breeding cycle. A little over three months later, on August 1, Lena had cubs of her own. If scarcity relates directly to value, then Lena's little cubs were three of the most treasured gems in the animal kingdom. Every Amur tiger born is literally priceless. Lena gave birth to a second litter in October 1998 and a third litter on July 31, 2000. She continues to be the most important female tiger in North America, and she is doing everything we could ever ask of her to perpetuate her kind. Sadly, tigers in Siberia continue to fall to poachers' guns, and the future of all tigers in the wild is bleak.

The Saint Louis Zoo has also been involved in high-profile animal rescues. In fact, two of my favorite animals here in St. Louis, grizzlies named Bert and Ernie, were rescued by Steve Bircher, our curator in charge of carnivores.

Grizzly bears are one of nine subspecies of brown bears, the species with the largest range of all the bears. The nine subspecies are spread across North America, Europe, and Asia, and of these nine, only two are thought to exist at present in North America—the grizzly (*Urus arctos horribillis*) and the Kodiak brown bear (*Urus arctos middendorffi*).

The Kodiak brown bear is found on the Kodiak Island chain, off the southwestern coast of Alaska. The grizzly bear is found in Alaska, Canada, and the continental United States. The majority of the estimated forty to fifty thousand grizzlies left in North America can be found in Alaska and Canada. Fewer than eight hundred are thought to remain in the Lower Forty-eight.

The silvery-tipped hairs of *horribillis* give it a grizzled appearance, hence the common name "grizzly." Grizzlies are omnivores, eating both plant material and animal flesh. Contrary to the popular belief, plant material, consisting of tubers, leaves, berries, and seeds, makes up the largest portion of their diet. However, as they are opportunistic eaters, they eat meat whenever it is available, usually in the form of carcasses of elk or moose that have died due to the stress of winter or advanced age. Grizzlies also love to fish. The bears get their fill of salmon during that fish's annual spawning run in summer and fall.

In the summer and fall, grizzlies can consume eighty to ninety pounds of food per day, gaining five to six pounds of fat per day. This prepares them physiologically for months of winter hibernation. In fact, the high level of stored body fat and the onset of wintry weather determine when grizzlies begin hibernation. This usually occurs in the month of November. The hibernating grizzly has a lowered heart and a lowered metabolic rate, while its body temperature remains near normal. This differentiates the grizzly from the truly deep hibernators, like woodchucks, ground squirrels, and bats, whose body temperatures can drop to thirty-eight degrees Fahrenheit. The grizzly does not eat, drink, urinate, or defecate during hibernation, but it can easily be roused to defend itself when disturbed.

Birthing occurs during this period, following a May or June mating. The spring-fertilized egg implants itself in October, and females give birth in January or February, giving a total gestation period between 230 days to 260 days. The average litter size is two, with a range of one to three cubs. Cubs weigh less than one pound at birth and reach a mature size of 300-800 pounds for males and 250-450 for females.

Bert and Ernie's story began in May 1990, when a family of three male cubs and their mother were living in the Chugach State Park near Anchorage, Alaska. The bears had wandered from the park into a residential neighborhood and had begun foraging in garbage cans and dog food dishes left on the porches of suburban homes. Despite repeated attempts to deter this behavior, wildlife protection officials had to destroy the mother after she was discovered feeding on the second-story balcony of a home.

Once a brown bear loses its natural fear of man and becomes habituated to this type of foraging behavior, a dangerous situation exists that almost always leads to the animal's death. Through educational programs and pamphlets, the Alaska Department of Fish and Game is constantly trying to inform the public of the dangers of encouraging unnatural bear feeding behavior. Typically, such encouragement is initiated by carelessly leaving excess food and garbage around homes, and this may attract bears. The department believes that if people living in bear country would keep their areas free of these temptations, some of these situations can be prevented.

When their mom was killed, the babies managed to get away, but they were soon back. They wandered into a large, fenced, vacant dog kennel, looking for uneaten dog food. Alaskans, by and large, are not easily intimidated by wildlife. The owner simply walked out and closed the gate to the kennel, trapping the bears inside. One captured cub was transferred to the Alaska Zoo for temporary holding until a permanent home was found at a zoo in North Dakota. The two remaining cubs, Bert and Ernie, stayed at the Alaska Zoo through the winter. By spring, the cubs weighed over one hundred pounds each and had outgrown their small quarters. Releasing them into the wild was not an option, according to Alaska Zoo officials, because they were so young when they lost their mother they'd never learn to forage for natural food.

Efforts to find a zoo home for the cubs in Canada failed, and a plea went out in May through the news media, asking if any zoo would take the bears. If no home could be found, the Alaska Department of Fish and Game decided the bears would have to be destroyed.

Brown bears are common in zoos today and are long-lived mammals, often reaching twenty-five to thirty years of age. Because they reproduce easily in captivity and have few wild places where they can be released, not many zoos have space for more. Fortunately, the Saint Louis Zoo, which has more bear species on display than any other zoo in the country, had a vacant grizzly bear exhibit. The last pair of grizzlies had died recently at the age of thirty.

Zoo officials read in the *St. Louis Post-Dispatch* about the plight of the bear cubs and immediately contacted the Alaska Zoo. Because of the publicity from the Associated Press and other news agencies, the Alaska Zoo also received numerous responses from conservation organizations and concerned citizens from all over the world. Alaska's Department of Fish and Game was delighted that the Saint Louis Zoo would take Bert and Ernie, who by now had become international celebrities.

Federal Express, the world's largest express transportation company, was contacted to help with the bear shipment. They generously donated both the shipment of an empty crate to Anchorage and the "fare" for the bear-filled crate (and our curator) nonstop from Anchorage to Memphis, and from there to Lambert International Airport in St. Louis.

The FedEx plane that brought Bert and Ernie to their new home at the Saint Louis Zoo.
SAINT LOUIS ZOO

Following standard zoo procedure, upon arrival the cubs were admitted to thirty-day quarantine at the Veterinary Hospital. Bert and Ernie earned a clean bill of health and were transferred to the grizzly enclosure in the Bear Pits.

At last, the day had come for the bears to make their debuts. On June 27 the news media was invited to the zoo to cover the introduction of the cubs to their new quarters and their adoring public. When the time came, Bert and Ernie slowly and cautiously came outside to inspect their new home. After a few minutes of smelling and viewing the surroundings, the bears started running and chasing each other, jumping into the pool and wrestling. They looked as if they enjoyed the exhibit immensely; maybe they knew their reprieve had come through.

Lions, and tigers, and bears—none are saved in zoos. They may be *safe* here, but they are not *saved* here. This may be the hardest thing to communicate to our audience. Animals are saved only if they can continue to live, relatively unmolested, in the wild. Zoos only provide temporary refuge for individuals, like Bert and Ernie, and for species, like the Amur tiger. But a refuge is not the same as a home. In the end, animals are only saved if they remain safe *in the wild* or can be safely reintroduced *in the wild*.

8

Golden Tamarins
and Ferrets with Black Feet

The current candidate for the dubious distinction of being North America's most endangered mammal might go to a charming little fellow called the black-footed ferret. They're members of the weasel family (Mustelidae), which includes minks, otters, wolverines, badgers, and skunks. They look much like the ferrets you might find in a pet shop, and both types of ferrets are related, but the ones in pet shops are a different species—one that evolved in Europe. Our North American ferrets look quite a bit like their European counterparts. They are bigger than a weasel, with a black mask across their face, a brownish head, a black-tipped tail, and black feet and legs.

The American naturalists John Audubon and James Bachman first described them in 1851, but they were not sighted again by Western scientists for another twenty-five years. Since they live underground and are nocturnal, that shouldn't be too surprising, but another reason they weren't spotted again for a quarter century is probably that there were never that many of them to begin with. They had a pretty big range, though, being found throughout the Great Plains from Texas all the way to Saskatchewan and east to the Dakotas.

In fact, they were found virtually everywhere that prairie dogs were found, for the simple reason that prairie dogs were pretty much all they ever ate. They would take over a prairie dog burrow and, every year in May or June, would give birth to one litter of four or five babies (called kits). The parents shared the hunting duties, slipping into nearby prairie dog burrows by night, killing an animal, and then dragging it back to their own burrow. The young kits could fend for themselves by the fall of the year and would move away to take over another burrow (after dispatching the current residents and eating them for dinner).

The ferrets were fine as long as there was a plentiful supply of prairie dogs, but by 1960, government eradication programs had reduced the original geographic distribution of prairie dogs by an almost incomprehensible amount— approximately 98 percent. The remaining populations of prairie dogs were too small and widely distributed to support the ferrets. To make matters worse, a disease called sylvatic plague wiped out many more prairie dogs, and to make matters even worse, the disease also claimed ferrets as victims. So the black-footed ferret was basically wiped out. During the 1970s, there were a few widely scattered sightings and, by the end of the decade, the only known viable population (which was in South Dakota) disappeared.

Then, in 1981, a dog killed an unusual animal on a ranch in Wyoming. The rancher took the animal to a taxidermist, who recognized what it was. Scientists flocked to the site, discovered a small population of black-footed ferrets, and began to carefully monitor the population. Between 1981 and 1984 the population of ferrets, undoubtedly the last ones alive, rose to 130. But things were about to turn against the ferrets one more time. In 1985 both canine distemper *and* sylvatic plague struck, leaving only eighteen ferrets alive. In 1987 the remaining eighteen animals were trapped out in a last-ditch effort to save the species through captive breeding.

At that point, the Conservation Breeding Specialist Group convened a meeting to plan the ferret's future. They decided to breed animals in multiple zoos (in order to protect the remaining survivors from a catastrophic loss due to disease) and then, hopefully, to begin reintroducing ferrets back to the wild in selected states. The Wyoming Game and Fish Department moved their remaining animals to the National Zoo's Conservation and Research Center in Front Royal, and later, in 1988, eight more ferrets were moved to the Henry Doorly Zoo. In 1990, the Cheyenne Mountain Zoo in Colorado Springs began breeding, followed by the Louisville Zoo and the Phoenix Zoo in 1991, and then the Toronto Zoo in 1993.

Progress was initially slow, but it was steady. The population grew from the eighteen originally left alive in the spring of 1987 to 349 ferrets by the fall of 1992. CBSG's original plan was to begin reintroductions back to the wild and, using computer modeling, they figured that it was safe to start the process once the population reached about 240 animals. Specifically, they wanted to have 90 males and 150 females of prime breeding age (one to three years old), *plus* they wanted a sperm bank, with sex cells from the most important males. They hit their goal by 1991, a year earlier than they had thought they could.

As zoos continued to breed ferrets, field scientists began to release animals back to the wild, first in Shirley Basin, Wyoming, in 1991, and then in the Russell National Wildlife Refuge in Montana and the Conata Basin in South Dakota in 1994. Another site was added in Arizona in 1996, in Montana in 1997, and along the Arizona-Utah border in 1999. By then, over 1,185 ferrets had been released back to the wild. To date, the original population of eighteen founder animals has resulted in over four thousand births. Not too bad for a small group of zoos!

In June 2002, CBSG convened a second meeting in Denver with members of the USFWS, the six zoos that were in the captive breeding program, and representatives from potential reintroduction sites. They had learned a good deal over the previous decade, but they still faced many problems. From the zoos' perspective, managing the ferrets and their genetics were the critical issues, while the USFWS focused on maintaining their natural habitat and balancing reintroduction by moving animals from one introduction site to another.

On one level, it sort of looks easy, but when you take a closer look, you see that there are many factors that zoos have to consider in order to effectively manage animals, things like medical problems, and a host of considerations related to genetics. For example, we have discovered that, in addition to canine distemper and plague, ferrets often get the same kind of respiratory illnesses, like cold and flu, that humans get. This means that keepers have to shower before entering the breeding facilities and wear freshly laundered clothes, clean shoes, and surgical masks when working around the animals. Because the population was so small to begin with, every animal has genetic importance. If we find two ferrets that don't get along, but the computer tells us that it's important for them to produce offspring, then we have to help them out a little. The way we help is to take sperm from the males, freeze it, and use it later to inseminate the females. The National Zoo has led all of the research in this area, and one of their reproductive biologists, Dr. JoGayle Howard, has developed a technique called laparoscopic intrauterine artificial insemination. She takes a slender instrument with a lens, called a laparoscope, and uses it to deposit semen directly into the uterus of the female.

Some of the most important issues the SSP must consider deal with basic demographics. For example, we could maintain a very young population, which would result in a higher number of births. But shortening the generation time also means that the population's genetic diversity will disappear more quickly. We also have found that ferrets that have been returned to the wild have better sperm quality. Now we're wondering if that is a result of genetics or of environmental factors such as diet. It could be that, for some unknown reason, ferrets that eat nothing but prairie dog carcasses produce better sperm!

We've also discovered, not too surprisingly, that ferrets that have access to "preconditioning pens" located outdoors and built with burrow systems occupied by real prairie dogs are ten times more likely to survive when reintroduced back to the wild. The problems for the people working on reintroductions, though, focus on the two issues noted earlier: first, disease continues to impact the reintroduced populations; and second, there just aren't enough prairie dogs in the wild to support self-sustaining populations of ferrets, except in one place, the Conata Basin.

To put the success of the Conata Basin program in perspective, in 2000 there were over sixty-five litters of wild-born ferrets, with well over 150 kits total. The four other most productive sites *combined* produced only twenty litters. This has led the collaborative of scientists working on the ferrets to conclude that the best option for increasing the ferret population is a program of *translocation*. Simply put, this means capturing wild-born ferrets from the original zoo stock and moving them to other sites. The real advantage is that the zoos involved in the program can focus on breeding smaller numbers of better-educated ferrets—that is, ferrets that have more experience hunting prairie dogs. Since the wild-born ferrets already have that experience, they should adapt to their new surroundings more easily.

In the end, the goal of the black-footed ferret program is simple. We want to improve the status of the species from endangered to threatened by establishing a minimum of ten free-ranging populations, spread over the widest possible area within their former range. Each of these populations should have a minimum of thirty breeding adults, and it is hoped that there will be a total of fifteen hundred free-ranging ferrets by the year 2010. So far, thanks in large measure to a small number of cooperating zoos, we are closing in on this goal. While the future of what was once North America's most endangered mammal is not yet secure, it is at least going in the right direction.

Few people are aware of the black-footed ferret and its close brush with extinction, but even fewer seem to be aware of the role that zoos played in another classic example of the role that breeding can play in rescuing an endangered species—the example of the golden lion tamarin. These beautiful little monkeys were once abundant along the Atlantic coastal forests of Brazil. In the hills above Rio de Janeiro you could hear their high-pitched whistles and, perhaps, see a brilliant flash of rich orange-gold as they leaped from tree to tree. But if you haven't been to Rio, you might still have seen them running loose around your local zoo. It turns out that one of the best ways to teach these beautiful monkeys survival skills is to set them loose on the grounds, and that's exactly what many zoos have done in an effort to prepare them to be returned to the wild. Currently, there are about 120 zoos worldwide housing golden lion tamarins, and some of these animals will find themselves leaving the zoos where they were born and living the remainder of their lives as citizens of the restored coastal rain forests of their ancestral home.

Their story begins in the 1960s, when a biologist named Adelmar Coimbra-Filho, the pioneer of Brazilian primatology, dedicated his life to saving three of the four different species of lion tamarins: the golden lion, black lion, and golden-headed lion tamarin. (The fourth species, the black-faced lion tamarin, was at that time not yet known to western science.) Of the four, the golden lion was the most visible and popular, but it suffered greatly from habitat loss in its home state of Rio de Janeiro, coupled with high demand from the pet trade. By 1968, the international zoo community, convinced that these beautiful little monkeys were headed for extinction, formally recognized them as endangered species and forced a ban on importation of tamarins to the United States. At the same time, the black lion was recognized as an even more critical species of concern. They were thought to be extinct for sixty-five years until, in 1970, they were rediscovered in the state of Sao Paulo.

The other thing that happened in 1970 was, from the tamarins' point of view, even more important. In that year a young man named Russ Mittermeier, who had just graduated with a BA from Dartmouth College, decided that he should go on for a Ph.D. *and* that he should do his degree on New World primates. Coimbra's groundbreaking studies had not really garnered much attention in the scientific community—not because they weren't good, but because they were all written in Portuguese! Mittermeier did not read Portuguese. But he did read Spanish, and he struggled through the translation. The material he read (or at least, the parts that he understood) enthralled him, and six months later in July 1971, he appeared on Coimbra's doorstep. As Russ says, "We struck up an immediate friendship that has

One of the four species of tamarins. CHUCK DRESNER

lasted for three decades." Together, Mittermeier and Coimbra crafted a plan to carry out field surveys in the Atlantic rain forest and to publish Coimbra's early work in English.

The international attention focused on tamarins helped Coimbra to work closely with the government of Brazil, and he was able to establish a breeding facility in Rio in 1973 and a reserve in 1974. But just having a single breeding facility and a reserve was not going to be enough. In 1972 and 1975, there were international conferences on tamarins, and in 1978 another conference was held, this time in Europe. Meanwhile, their early plan—to continue to survey the rain forest for every primate species they could find—bore fruit. Beginning in 1973 and running more or less continuously to 1985, studies were conducted that had the net effect of forcing the world to recognize the Atlantic forest as one of the biodiversity hotspots of the world.

At the same time, a young woman named Devra Kleiman was organizing the international zoo community to help bring the tamarins back from the brink of extinction. Devra got her Ph.D. from the University of London in 1969 and went to work at the Zoological Society of London, where she worked with rodents, bats, canids (members of the dog family), and weasels. In 1972, she came to America to work at the National Zoo in Washington. She got involved with tamarins immediately. While there was already an inventory of animals housed in the United States (in the form of a studbook), there had never really been any systematic research into how best to manage the population. Once Devra took over, she began to actively manage breeding, change diets, and create more effective social groups for the tamarins. She was so successful that, by the end of the 1970s, the National Zoo had a surplus of animals and could begin a program of loaning captive-bred tamarins to zoos throughout the United States.

In 1981, under Devra's sure hand, the National Zoo stepped in and crafted an agreement with the government of Brazil to develop an integrated program of research and conservation both in the reserves and in zoos. Devra created the first zoo-based management committee for the golden lion tamarin. Management committees for other species, modeled on Devra's groundbreaking efforts, soon followed, and in 1984 the National Zoo began reintroducing captive-born tamarins onto private lands.

With the completion of the field surveys in 1985 under Russ Mittermeier, who was then working with the World Wildlife Fund (he's now with Conservation International), the conservation community was ready to take the next step. In 1990, CBSG and the irrepressible Ulie Seal convened a meeting of all of the interested researchers, government officials, zoo authorities, and major landowners to craft a long-term conservation plan. This plan was designed to focus research efforts both in zoos and in the field, establish breeding programs, increase the size of wild populations, enhance the genetic diversity of those populations, develop additional protected areas, and finally, develop a public education program primarily focused on helping Brazilians to value their tamarins and, more importantly, to preserve their habitat.

While there are now different programs in place for all four species of tamarins, the program developed for the golden lion tamarin in Rio de Janeiro is probably the best example. First, it is a model because, from its inception, it combined the expertise of zoo researchers and government officials and collaborated with Brazilian scientists working at the Centro de Primatologia do Rio de Janeiro in a coordinated effort. It has already trained numerous Brazilians and has done a remarkable job in public education. It has reintroduced captive-born tamarins to the wilds and has restocked private forests near the reserve, involving local people in conservation and effectively increasing the protected area. Even more importantly, it has resulted in expanded habitat by promoting reforestation, and most recently the creation of forest corridors between forest fragments and the reserve. In a largely symbolic but critically important move, the collaboration invited the 120 zoos that held captive tamarins to return the ownership of the animals to Brazil. Even though we have them here in St. Louis, they belong to their nation of origin. Thanks to the continued passion of Devra Kleiman and the National Zoo, the collaboration formed an International Management Committee to advise and manage both the captive and the wild population. This is critical because the tamarins still live in fragmented, noncontinuous forests. Inbreeding is a perennial problem. By tracking the family trees of both wild and captive populations and then either introducing captive-bred stock into the small populations or translocating animals from one patch of forest to another, the population as a whole can maintain its genetic diversity. Finally, if you asked Devra, she would say that the thing she is most proud of is that the tamarin project has become a truly Brazilian initiative. Over the years, this project has trained a whole generation of Brazilian scientists and conservationists. Not all of them have worked strictly with tamarins, but all of them have contributed to the understanding of the remarkable Atlantic forest habitat, a habitat that might not even exist today, were it not for her efforts and the efforts of a handful of other dedicated pioneers.

This marvelous collaborative effort has set for itself some very important long-term goals we feel must be achieved over the next several years. In the "Countdown to 2025" effort, created by the National Zoo, the collaborative would like to expand the amount of forest habitat four-fold, to about sixty thousand acres. Currently there are about twelve hundred golden lion tamarins in the wild. The collaborative would like to see that number doubled. They would like to have the population genetically diverse, and they want to broaden their training programs and increase their education efforts for locals. Finally, they would like to use the model developed for the golden lion tamarin to help save the other three species of tamarins. The extinction clock is still ticking for the tamarins, but it has slowed. Thanks to a remarkable collaboration, tamarins have been rescued from extinction. Now we must do more—now we must secure their future.

Today there are quite a few reintroduction programs, many of them involving US zoos and some of them involving species from the United States. Some, like the reintroduction of the California condor and the red and gray wolves, are highly visible and, at least in the latter case, somewhat controversial. Others, like the reintroduction of endangered invertebrates such as butterflies and beetles, are largely overlooked by the general public. I'm glad that people follow the success projects like the California condor with such great interest, but I guess I'm disappointed that our work with animals like the Channel Island fox, the Karner blue butterfly, and the American burying beetle goes largely unnoticed.

I'm especially concerned because, while it might be comparatively easy to get people interested in animals like the Channel Island fox, with its beautiful eyes and soft coat, it's another matter entirely to get people excited about beetles, or even butterflies. This, despite the fact that the beetles we work with are among the most fascinating creatures on earth, and the butterflies we are saving—flying flowers that count their time not in months but in moments—are among the most dazzling creatures in all of creation.

So let me tell you about beetles and butterflies. I'll begin with the American burying beetle, not so much because it can compare with a butterfly's beauty (though it is a stunner), but because its natural history reminds me of something from a Stephen King novel.

The American burying beetle, *Nicrophorus americanus,* is about two inches long and has a black, shiny carapace (or armor shell) with bright orange markings on its back, wing coverings, and head. It also has strong pincers, used for fighting and ripping through food, and incredibly sensitive antennae that can pick up a whiff of a mouse or quail carcass within an hour of death and from as far away as two miles!

Strong fliers, the beetles converge on the dead animal, and the battle is on. There are rules that circumscribe the fight for the carrion—males fight males and females fight females. Usually the biggest, strongest male and female triumph, driving the other beetles away. They are now a couple and they can begin the process of raising a family of their own.

Their first order of business is to find a suitable place to bury their dead trophy. They slide under the body, turn over on their backs, and use their powerful legs to move the carcass—sometimes relocating it up to one hundred yards away. They then dig furiously, excavating the soil beneath the body and mounding it around until, several hours later, it is completely buried. Next, working underground, they roll the carrion into a ball and embalm the corpse using secretions from their mouths and anuses that strip away the fur or feathers of the mouse or bird and help preserve the corpse in a semi-mummified state. They mate and the female can now lay her eggs in the soft soil around the preserved body where, within a few days, they will hatch.

The larvae then beg for food from their parents, who rip flesh from the embalmed corpse and regurgitate it. The larvae beg by tickling mom and dad's jaws. This goes on for about a week, until the young have eaten all but the bones. The adults then fly away, and the young pupate in the soft soil, emerging about a month later as adults in their own right. Brood sizes range from one to thirty young, but twelve to fifteen is closer to the average. Interestingly, it appears that mom and dad make sure that they have enough food by eating any excess larvae, ensuring that their remaining offspring will have enough food to grow.

Ideally, the carcass is buried at nighttime because this prevents flies, which are active during the day, from laying eggs of their own on the carcass. The beetles also have a symbiotic relationship with tiny mites that live on their bodies. The mites jump off the beetles' bodies and eat whatever eggs are deposited by flies. The adults must continually tend the carcass, removing any fungus that starts to grow and covering the carrion with a special antibacterial secretion. These secretions not only preserve the dead body, but also seem to slow decomposition as well, allowing plenty of time for the larvae to feed.

The burying beetle is not alone among insects that provide parental care to their young. Bees, wasps, ants, and termites do much the same thing. The difference is that all of those insects are *social* insects. They live in huge colonies or hives. The burying beetle, amazingly, acts more like a human couple than it does a social insect (except for the part about eating their excess young).

The American burying beetle was once found throughout the eastern and central United States (thirty-five states in all), plus southern Ontario, Quebec, and Nova Scotia in Canada. While we're not sure exactly when their numbers began to decline, they are now found only on Block Island, off the coast of Rhode Island, and in tiny patches located in seven other areas in the midwestern United States. The reason for their decline is even more mysterious. One theory is that the decline began with the extinction of the passenger pigeon, a bird that would have probably been close to the ideal size for a hungry mom and dad looking for a suitable host for breeding. A more complex theory says that, years ago, the top station in the food chain was occupied by big predators that are no longer around in significant numbers—wolves, bears, and big cats. As hunting and the conversion of fields and forests to agriculture continued, the top predators declined, and their position was filled by what we might call second-tier predators, such as foxes, raccoons, and skunks. These animals focus their attention on killing and eating the smaller animals, such as birds and mice, that the burying beetle prefers. Overuse of pesticides may be a contributing factor, but the beetle's decline started before the use of DDT became widespread. Habitat fragmentation could also be a culprit, but there could be any number of other contributors. For example, because they do their best work at night, perhaps something like light pollution is a factor. Who knows? Or perhaps disease played a role.

The greater questions, I suppose, are "What are we doing about it, and why should we care?" Zoos began to work on the American burying beetle in 1995, when Tony Vecchio, the innovative and progressive director of the Roger Williams Park Zoo in Rhode Island (he's now at the Oregon Zoo in Portland) and his staff, led by Dr. Lisa Dabek, figured out how to breed the beetles in captivity and began a release program. Working with the USFWS, the zoo released over one thousand pairs of beetles in nearby Nantucket, Massachusetts, and has continued to release beetles on annual basis, studying their progress all the while. More importantly, they have helped other zoos, like the Columbus Zoo in Ohio and our own Saint Louis Zoo, to develop programs for breeding and reintroduction in other states. In fact, the American burying beetle has become one of the twelve major projects in our zoo's new WildCare Institute, a project that I'll describe in more detail later on.

The harder question to answer is, "Why should we care?" Since the landing of the Pilgrims in 1620, not far from where the Roger Williams Park Zoo started its battle to save the burying beetle, more than five hundred species, subspecies, and varieties of our nation's plants and animals are known to have become extinct. What's one more to the list? I suppose there's at least one practical reason: because they are carrion feeders, the burying beetles ultimately return valuable nutrients to the soil. There may, however, be other practical reasons. We know next to nothing about the antibiotics that the beetles produce naturally. We know next to nothing about how they can sense the presence of a corpse over such a vast distance. In fact, we don't know much about these guys at all, and some of what we *don't* know could be of tremendous help to our own species. Studying them is also important because, like amphibians, they are probably a pretty good indicator species—one that tells us if the overall environment is healthy. Understanding why they disappeared might help us to better understand general environmental issues that may be of critical import to humans. In the end though, I like World Wildlife Fund's reason the best: "All that lives beneath Earth's fragile canopy is, in some elemental fashion, related. Is born, moves, feeds, reproduces, dies. Tiger and turtle dove; each tiny flower and homely frog; the running child, father to the man and, in ways as yet unknown, brother to the salamander. If mankind continues to allow whole species to perish, when does their peril also become our own?"

It is easier to justify our work with another class of invertebrates—butterflies—on purely aesthetic grounds. They are, simply put, beautiful, and it is easy to see how our world will be made poorer without them. Butterfly conservation in American zoos became an organized initiative in 2001 as a result of several different forces. First, individual zoos were having a difficult time trying to deal with USFWS on an individual basis. Not surprisingly, Fish and Wildlife preferred an organized, integrated effort as opposed to having to deal with many zoos in many different states. Perhaps even more importantly, smaller zoos were coming to the conclusion that the big guys, like Saint Louis, Chicago, San Diego, and the Wildlife Conservation Society, were getting all the attention and, more importantly, they were getting the attention more for what they were doing in far-flung parts of the world than they were for working closer to home.

So, in the fall of 2001, the American Zoo and Aquarium Association organized the BFCI, or Butterfly Conservation Initiative, which now consists of forty-seven participating zoos and five very important partners. The key partner is, of course, the USFWS, largely because the BFCI is designed to support the recovery of the twenty-two federally listed species of threatened, endangered, or vulnerable butterflies. The BFCI is popular with zoos of all sizes, but for several reasons it is an ideal program for smaller zoos. First, it really doesn't take much space or particularly specialized facilities to breed butterflies. Second, it's easy to find local partners, like The Nature Conservancy, that have tracts of land that might make suitable reintroduction sites. Third, it's easy to monitor success at release sites—virtually anyone, including volunteers, can be trained to do it. Finally, butterflies are popular in educational programming. People love them, and it's easy for zoos to develop programs for animals that people already appreciate.

The first program focused on the Karner blue and began in June 2002, when sixty-seven wildlife biologists, university scientists, private landowners, and zoo professionals met at the Toledo Zoo. There were twenty zoos represented at the first meeting, and they, along with the USFWS, began programs in several states. Within months of the meeting, reintroduction programs started in New York, New Hampshire, and Michigan. Much of the information needed by the partners was already available from the folks at the Toledo Zoo, who had been breeding and releasing butterflies in Ohio for years.

The success of the Karner blue effort led to other successes. A similar partnership was developed for endangered species of California butterflies and butterflies of the Northwest Coast, including the Fender's blue, Puget blue, and Oregon silverspot. A fourth program was developed in Florida for the Miami blue. This butterfly, confined to a small area near Miami, was thought to be extinct after hurricane Andrew swept through the area in 1992. In 1999, a small remnant population was found, and in 2003, one of the BFCI's key partners, the McGuire Center for Lepidoptera Research, began to rear the Miami blue in captivity. In May 2004, five hundred Miami blue caterpillars were released in two of Florida's National Parks. Not bad, considering that there were only a total of fifty adults left alive in the wild by that point.

While the Saint Louis Zoo has not been active in butterfly conservation, we have been cooperating in a fascinating program with a charismatic but little-known mammal called the Channel Island fox. Dr. Cheryl Asa, who heads our reproductive science efforts here at the zoo, has keyed this work. The Saint Louis Zoo is the home of our nation's center for the study of contraception in wild animals, but Dr. Asa has always been interested as much in reproduction as she has been in contraception (and the two do go together pretty much hand-in-hand). She shared with me the following information.

The Channel Islands can be seen on the Pacific horizon from the grounds of the Santa Barbara Zoo, an institution that was among the first to work with the Channel Island foxes. (In fact, most of what we know about rearing Channel Island foxes comes from the zoo in Santa Barbara). These foxes are slightly smaller cousins of the gray fox that is found on the US mainland. They diverged from the mainland species thousands of years ago, and now the fox populations on each of the Channel Islands off the coast of southern California are considered separate subspecies. This taxonomic distinction is at the heart of endangered species designation and recovery. Mainland gray foxes are plentiful, but because each of the Channel Islands is small, the fox population on each island is also small, and thus more vulnerable to threats. In fact, the assortment of threats facing these unique little animals provides a lesson in miniature of how ecosystem changes can lead to extinction.

The fox population on San Miguel Island first crashed about seven years ago. When estimates were down to about thirty animals, the National Park Service, the government agency that oversees several of the islands, decided there was no choice but to capture the remaining foxes and bring them into what amounted to "protective custody." They assembled a recovery team made up of fox ecologists, geneticists, population biologists, veterinarians, and several of us representing zoos, since we could provide special knowledge about fox captive husbandry and reproduction. They felt that if the foxes were going to be held in captivity while the biologists tried to determine the cause of the crash, it would be best if they reproduced and increased their numbers for possible future release.

The bad news was that by the time trapping began, only fifteen foxes were found, and fourteen of those were captured. The further bad news was that of the fourteen, only four were males; now, that wouldn't be a problem if this was a polygamous species, but foxes, like many other canids, are monogamous. This means that only four breeding pairs could be formed that first year, and only two pups were born. The good news is that all those captured adults survived, and each year the population continues to increase. Dr. Joan Bauman, our endocrinologist, is measuring reproductive hormones in fox feces to help us monitor breeding condition.

Meanwhile, the ecologists and vets are zeroing in on what caused the crash, which turns out to be a rather convoluted tale starting with ranchers who had sheep and pigs on the islands in the early part of the last century. When they left the island, they were able to capture all their sheep, but not all the pigs, and the pigs that were left successfully carved out a niche for themselves that has spelled disaster for the foxes. Golden eagles entered the picture, attracted during the breeding season by piglets, which were an easy meal. Once accustomed to finding prey on the island, they shifted their attention to foxes when the piglets got too big to catch and carry easily. The ecologists think that some eagles then began nesting on the islands, bringing even more predation pressure to the fox population, which may not have ever had any natural predators.

So, the answer is to find and remove the feral pigs as well as the golden eagles. To help convince the golden eagles not to come back, bald eagles are being reintroduced to the islands. Bald eagles eat fish, not foxes, so this may work!

As mentioned earlier, the other islands face their own problems. Canine distemper, brought onto Santa Catalina Island in the south by domestic dogs coming with their vacationing owners, has killed a substantial portion of that fox population. Even more complicated is the situation of competing endangered species on San Clemente Island, also in the south, where the island's foxes were found eating the eggs of the San Clemente shrike, an endangered bird species. The recovery program has to try to balance their competing interests. The most ingenious approach used so far is the "invisible fence" training collars placed on foxes, with the "fence" encircling the most popular shrike nesting area. It seems to be working (at least better than it does with my bulldogs, who occasionally decide that whatever is on the outside of the fence is worth a small shock to get at). The problems aren't completely solved, and we don't know what kind of threats may hit the other islands, but we are making headway.

Not all reintroductions involve putting animals back into the wild. For the last decade or so, the Saint Louis Zoo has been involved with reintroducing animals back into social groups here at the zoo. In particular, we have been reintroducing chimpanzees into social groups. You could call it the creation of a big foster family. Chimps normally live in groups that form around families. In the wild, those groups periodically come together into larger groups and then break apart into smaller groups (primatologists call these "fission/fusion groups"). There is a strict dominance hierarchy, with the alpha male running the show, at least in theory, and the alpha female playing a very important, but less dominant, role in the social group.

Their social lives are fluid and dynamic. Chimps are constantly forming alliances and looking to advance in their tight little social group. One or two lower-ranking males may seek to upset the top male by ganging up on him. Their daily political lives make our extended presidential primaries look tame. Dominance is not just measured by who has first access to food or a coveted place to sit—it also involves sexual access. Normally, the dominant male would have first access to a female in heat. Among chimps, though, secretive sexual liaisons are common. This means that a lower-ranking male might seek to one-up a higher-ranking male by getting to a female in estrus when the dominant male is otherwise occupied. Females certainly also show preferences for whom they mate with, and male dominance is clearly not the only factor in who mates with whom.

It's hard enough to manage a group of chimps that are related and have grown up together. It's even harder, though, to bring together a group of unrelated males and females of different ages and try to form a normal social group. Of the nine chimps in our group, only three are biologically related: Smoke and Mollie have a daughter named Cinder (a chimp we'll meet again later). The rest of the chimps are unrelated, but even an expert probably couldn't tell just by observing their behavior. Late in August 2003 we began to introduce a new chimp into our social group. At the time, she was just a baby, twenty-two months old and weighing twenty-two pounds. Now, over a year later, the process still isn't complete, but the baby, named Tammy, is making remarkable progress.

Tammy's story, at least from our point of view, began when a vet from the North Carolina Zoo was asked to come down to the Bayamon Zoo in Puerto Rico to treat a sick hippo. While at the zoo, the vet noticed a little baby chimp, which was being reared by hand at the zoo. Tammy had been pulled from her mother, who, at least ostensibly, was not caring for her infant (something of an oddity, given that the mother, a chimp named Katrina, had successfully reared babies in the past). We later found out that Katrina was on display in a small cage with no privacy, and when she went into labor, someone called the press in. I can only imagine what it was like with people gawking and camera flashes going off. In any case, Katrina abandoned her baby, and after eight hours, the staff of the zoo pulled the infant for hand-rearing. (Katrina died a year later under unusual circumstances—our vet saw the necropsy report and noticed a high level of pesticide in her blood but was not able to determine if that had anything to do with her condition. Luckily, the zoo now has changed veterinarians and directors, and these people are trying, under difficult circumstances, to improve the lives of these animals.)

Tammy spent the first year of her life under the watchful eye of a keeper named Ernesto Marquez. Ernesto tried his best to raise Tammy appropriately. That meant interacting with her much as a mother chimp would interact with a baby. In addition to carrying the baby and not leaving it alone, the keeper must make appropriate chimp gestures and vocalizations. At the same time, baby chimpanzees need an incredible amount of positive feedback in order to develop as normally as possible. They need almost constant nurturing and physical contact, just as their chimp mothers would provide, plus they need an appropriate environment in which to develop their physical skills.

While Ernesto clearly did many of these things, he also occasionally dressed up Tammy and took her around the zoo grounds—not a very appropriate behavior. The vet from the North Carolina Zoo brought all this to the attention of Lorraine Smith, who worked at the same zoo. Lorraine, in turn, called the USDA inspector, Dr. Sylvia Taylor, and she recommended that Tammy be transferred to another zoo. Because of our experience in building chimpanzee "foster families," the SSP selected the Saint Louis Zoo as the best choice for Tammy.

Tammy wasn't reared in complete isolation from other chimps. In fact, her biological father, a chimp named Yuyo, lived in an adjacent cage and often played with her through the mesh that separated them. Yuyo had a story of his own. He was thirty years old at the time, but he hadn't lived in the zoo for his entire life. For part of his life, he lived at another zoo in Puerto Rico, and for ten of his thirty years he lived on his own after escaping. In Puerto Rico, he had lived off the land (and by raiding trash cans) and is probably why the myth of the Chupacabera started, a local legend about a monster that kills animals.

Ernesto was genuinely torn when two members of our staff showed up to bring Tammy to St. Louis. On one hand, he had always wanted Tammy to grow up in a normal family environment. On the other hand, he wanted to keep Tammy—he was deeply attached to her. He didn't have much to say about it, though. The zoo had gotten a new director after Katrina died (and the vet who cared for her was fired), and Ernesto lost his job as well. With his departure, the job of caring for Tammy was passed on to two young women, Adlim Maldanado and Lourdes Mendez. Lourdes is a biologist and a paid enrichment coordinator, and Adlim was working as an unpaid volunteer—her sister Enid helped, too. They took Tammy's diapers off of her and stopped dressing her in human clothes and, more importantly, followed Lorraine Smith's advice on rearing techniques. Still, we were acutely aware of the possibilities of Ernesto resisting the USDA's order to move the animal by stirring up the local press. Tammy was already the darling of the community, so the situation was even more volatile. On the other hand, the two people we selected to go down and get Tammy were absolutely perfect for the job.

One was Terri Hunnicutt who, in addition to having an enormous amount of experience with chimps, also happened to be about the same age as the two volunteer keepers who were rearing Tammy. They bonded more or less instantly when they saw how Terri interacted with the little infant. The other person we sent down was one of our vets, Dr. Luis Padilla. Luis just happened to hail from the same area of Puerto Rico, so in addition to speaking perfect Spanish, he had the advantage of being sort of the local boy who made good.

Terri and Luis spent four days getting to know Tammy and threading themselves through the politics of the local community. They divided their time between tours of the town, luncheons and dinners, and meetings with the press, and bonding with the infant chimp. Terri, in particular, was horribly frustrated by all the formalities, in part because she had to deal with the ever-present specter of Ernesto, who constantly called the zoo and occasionally showed up in person. Clearly the two young keepers were terrified of him. At the same time, Tammy was besieged by visitors. It seemed that everyone in the town knew her and had to come by to see how things were progressing.

Tammy had a small cage, but every day they would let her loose in the vet clinic. She would go into the offices, grabbing the phones and scattering papers everywhere. Terri painted this picture by asking me to imagine what it would be like to let a young chimp run loose in my office. If you could see the stacks of paper I have mounded all over this room, you could envision the havoc that would ensue. Occasionally, Tammy would escape from the vet offices. The offices were blocked off from the rest of the zoo by gates, but Tammy could easily scale them and take off, and the girls couldn't fit through the gates to run after her. Terri, on the other hand, is tiny, and she could easily slip through to chase the chimp. She soon discovered that, while she wasn't fast enough to catch Tammy, she could outwit her by pretending that she was going to leave her by herself; eventually, Tammy always came back to her.

By the fourth day, Terri was almost exhausted. She'd been sleeping with Tammy every night, and between caring for the baby and helping Luis to run interference with the press and the local dignitaries, she was emotionally about spent. I think the last straw was when the zoo director came by to tell her that the press would be in at five o'clock the next morning to interview her and cover the process of loading Tammy into the truck that would take her to the airport.

Luis gave Tammy a small dose of valium in the morning, and Tammy was fine until they reached the airport. Even as they lifted her little crate up to the cargo hold of the FedEx plane, she seemed calm although, as Terri says, "Her eyes were as big as saucers." Terri and Luis checked on Tammy periodically on the flight to the FedEx hub in Memphis, where our staff was waiting to drive everyone on to St. Louis. I met Tammy shortly after their arrival here at the zoo, and like everyone who has ever come in contact with the little darling, I fell instantly in love.

We kept Tammy in quarantine for thirty days and then began the process of integrating her with her new chimp family. During the quarantine period, Tammy was with one of our keepers twenty-four hours a day. Ingrid Porton, our curator of primates, and two of our keepers (Terri and Lauren Marvel) spent most of the time with her. After her quarantine, Terri carried Tammy over to her new home and, with all the chimps out of the holding area, allowed Tammy to get used to the new enclosures.

They decided to begin the integration process by allowing Tammy to bond with an eleven-year-old female named Mlinzi. Mlinzi was an obvious choice, since the only other chimp that could rear Tammy (Mollie) still had an eight-year-old infant (Cinder), who would probably be jealous of the new arrival. For two days the keepers stayed with Tammy while Mlinzi watched from an adjacent enclosure. After two days, the keepers began the process of slowly backing out of the cage. The goal was to have Tammy alone in the cage, devoting all of her attention to Mlinzi and not to her human companions. Eventually we were successful. At the first sign of positive interaction between Tammy and Mlinzi, the keepers, now out of the cage, opened the shift door that separated the two enclosures. Mlinzi immediately ran in and hugged the little baby. Tammy, at least initially, wanted nothing to do with her. Within a few minutes, though, Tammy settled down, and Mlinzi began to do just what she would do with a baby of her own, gently grooming her while holding her closely with her free arm.

We are still in the process of integrating Tammy with our whole troop. She has a good deal to learn about being a chimp in a social group, and such learning

Cinder, our chimp with alopecia, and Tammy, the newest chimp to be introduced into our foster family. CAROL WEERTS

can only come over time. In particular, she needs to understand when she should stay and play or cut and run. Male chimps can put on quite a display as they vie with each other to maintain their place in the social order. Youngsters have to learn that the smart thing to do is simply get out of the way.

But Tammy is a remarkable creature. According to Terri, the person who perhaps knows her best, she has "grace, courage, and adaptability." Her little mind is constantly working, and she is building her physical skills. Terri explains, "She's not just reacting to her social environment, she's also working to make her own place in it." As I write this, she still is not fully integrated with all of our chimps. She has met our oldest male and is getting to know Jimmy, the chimp that is vying with our dominant male to be our troop's leader. With time, she will look and act as if she had been with these chimps all of her life.

We have discovered, after working with many species of primates, that reintroducing animals that rely heavily on learning (as opposed to instinct) to the wild, or even to other animals of the same species, is a complex undertaking. But what we learn from introducing an animal like Tammy to a new social group may ultimately help us better understand how to introduce chimps back into social groups in the wild. As every zoo professional will tell you, it is far better to save animals in the wild than it is to save them in zoos and later attempt to reintroduce them to the wild. Think of it this way: it's darn hard to get the toothpaste back in the tube.

9

Ralph Neds I Have Known

Okay, the chapter title requires some explanation. If you look closely as you wander around any zoo in America, you'll see that most keepers have radios. In fact, here at the Saint Louis Zoo, we have about two hundred radios in use on any given day. The advantage (and disadvantage) of radios is that anything that is broadcast can be heard by anyone standing anywhere near a person with a radio. This can be a problem in the case, for example, of an animal escape. Imagine the pandemonium that would ensue if you overheard a message that there's a cheetah loose in our zoo's River's Edge exhibit. It would be far easier to respond appropriately without having to deal with panicking visitors. The solution we used at the Indianapolis Zoo would be to (calmly) broadcast the message in code: "We have a Ralph Ned cheetah in the River's Edge." Anyone overhearing that would likely think, at worst, that we had a difficult time selecting a name for that particular animal.

The example of a loose cheetah in the River's Edge exhibit is not one that I just made up. In fact, twice we've had a cheetah get out of the exhibit in the River's Edge. As it happens, if you had to pick an animal that you'd want to escape (not that you want *any* of them to escape), it would have to be a cheetah. When we go into their enclosure for routine activities, the only precaution we take is to carry a big stick. I'm not precisely sure why we carry a big stick, but we do. In any case, cheetahs are among the least aggressive of any cats, and they're relatively harmless.

Both of our cheetah escapes took place under unusual circumstances, but both had a similar root cause—sex. Both times we introduced a male to a female, once in the main exhibit, and on the other occasion in our holding area, and both times the female wanted nothing to do with the male.

These weren't the only times that the Saint Louis Zoo had a big cat escape. Steve Bircher, the curator of carnivores, recalls that after he had been at the zoo for only a few years, he made an enormous mental error. He simply left the door of the lion holding area open and left the lion's food sitting outside the door. He then let the lions into the building, never having closed the door to the cage. To this day, he can't for the life of him figure out what he was thinking.

He opened the door to the outside exhibit and, as always, the big male was the first to come in. He went through the first shift cage and was moving through the second and toward the third when Steve saw that the door from the third cage into the keeper area was wide open. Before he could drop a shift door, trapping the male in the middle cage, the big guy was already through. The lion immediately noticed the door was open and that his bucket of food was standing about two feet away from Steve. Steve was numb with terror. The lion began to growl menacingly, although Steve never was sure if the lion was growling at him or growling because he was close to his food. I guess, either way, it didn't look good for Steve.

Steve began slowly to back toward the exit doors and, when he felt them behind him, he let himself out. He immediately called the senior animal division personnel who, as luck would have it, were all meeting in Charlie Hoessle's office. Bill Boever, now the zoo's director and chief operating officer, but then the zoo's vet, brought tranquilizing equipment, and Charlie and Roger Birkel, another long-time staff member, brought two .44s. Their plan was to cut through the roof of the building; not knowing exactly where the lion would be, they would then drop into the inside of one of the empty cages.

But the lion made things easy for them. As soon as they dropped into the building, the lion backed into his cage and accidentally banged the door shut, latching it securely. He had already eaten the food that Steve had left outside his cage and apparently had little motive to explore. Steve was called into the director's office, who at that time was William Hoff. Mr. Hoff was a former accountant, and it has been said he had an accountant's personality. "Young man," he said, looking over his glasses at the terrified young keeper, "in my old job, if we made one mistake, we were immediately fired." "Here at the Saint Louis Zoo," he continued, "we let you make one mistake and don't fire you until the second one." "You," he finished, "have made your one mistake."

Charlie was easier on Steve. He said, "You've joined the Jerry McNeal club." Steve immediately announced that the Jerry McNeal club was most emphatically not a club he wanted to be a member of. Charlie just grinned and said, "Yeah, but I guarantee you that you'll never make a mistake like that again for the rest of your career." So far, Charlie's prediction has held true.

Of course, it's not just Steve who doesn't want to be a part of the Jerry McNeal club. *Nobody* working here wants to be in the Jerry McNeal club, including Jerry McNeal. The club was formed on October 6, 1970, when the zoo finally recaptured a three-and-a-half-foot spitting cobra (or Ringhal's cobra) that had been loose in the Reptile House for forty days. Jerry McNeal was the man who accidentally let the snake out. He still works as a keeper in the herpetological section of the zoo and, twenty-four years after he let the cobra out, he still gets a little embarrassed over the whole thing.

Jerry's misadventure started when he was doing a routine cleaning of the cobra's enclosure. There was a small drainpipe in the exhibit that allowed excess water to flow down to an open trench that ran along under all of the exhibit tanks and fed into a sewer. The drain was only an inch and a half across and was kept sealed with a rubber plug. Jerry removed the plug and began to clean the exhibit when a radio call came saying that there might be feral dogs in the zoo. Back then, there wasn't a fence around our zoo (now a requirement) and sometimes wild dogs would roam onto the grounds, scaring animals and visitors alike.

Jerry was part of the team that went out to investigate. The problem was he neglected to put the plug in the drain before he left. When he returned, the cobra was gone. Jerry immediately wrote a Closed for Repairs sign, moved the public out of the Reptile House, and taped the sign to the front door. I have a photo of the sign. It actually looks pretty good (considering his hands were shaking when he did it).

His hands were shaking because the Reptile House, built in 1927, had numerous radiator pipes, drain lines, crawl spaces, ducts, and passageways. It was sort of a paradise for a snake that didn't want to be found, and spitting cobras tend to be reclusive by nature. Despite their shyness, they are dangerous. Spitting cobras tend to go for their victim's face, but the venom is not likely to cause blindness if it is quickly rinsed out. Fortunately, their fangs are not too long, and a heavy pair of pants would probably provide a fair amount of protection against a bite. At that point, though, Jerry was particularly worried about what his bosses would say.

Jerry immediately called Charlie, who was at that time a curator, and Ron Goellner, who was then a keeper. They had the most experience with venomous snakes. They started by barricading every possible nook and cranny within the Reptile House, and then they conducted a very thorough search of the building.

Finding nothing, they searched the grounds outside the building, including the trench the drain emptied into, and then searched the building next door. Again finding nothing, they concluded that the snake was still probably in the building. The keeper records showed that the cobra was about ready to shed, and snakes are normally pretty hungry once a shed is complete, so they developed a two-pronged strategy for catching the snake. First, they set out thirty hooks, baited with mice, and second, they spread talcum powder over every flat surface. Then, every morning before Jerry and the other keepers came to work, Charlie and Ron would come in early to check for any sign that the snake had slithered past or, even better, had latched onto one of the baited hooks. They did the same thing every evening after the staff left.

Mr. Hoff explained to the press that if the cobra did swallow one of the mice, it would probably kill it. "Of course," he said, "we'll try to operate to try to save it, but it's more important that we get the snake back—dead or alive." He also warned zoo visitors that if they spotted the cobra outside the Reptile House, they should stay at least ten feet away, call for help, but keep the cobra in sight until zoo keepers could arrive to trap it. (Liability laws have changed so much since then!) Oddly, attendance shot up during the time the cobra was missing. I guess it added a certain allure to the visitor experience.

In retrospect, I have to feel more sorry for the staff than I do for Ron and Charlie. The two men sort of enjoyed the snake-search, even if it did mean extra hours added to their day. But at least they were constantly on guard and, in theory, well prepared for a chance encounter. The staff members, on the other hand, were trying to do their regular jobs—the whole while keeping an eye out for a longish, reddish-brown killer.

On the morning of October 6, while they were hunting for the snake, Charlie and Ron finally lucked out. The snake had never gone for any of the baited mice and, in fact, had never traveled far from its exhibit. As they were crawling through a shallow crawl space inside the Reptile House forty feet long and eight feet wide, Charlie and Ron spotted their quarry.

After forty days of doing the rounds twice a day, the whole thing had become a little dreary, and Charlie and Ron had sort of given up hope. As it happens, when they finally spotted the snake, they didn't have anything to catch it with. They'd given up dragging all that stuff around with them days before. They did, however, have a flashlight, so Charlie shone the light in the snake's eyes and Ron ran for grabbers. He came back a few minutes later, breathless and armed with the four-foot snake hooks. Charlie just looked at him. "What?" said Ron. "Eye-shields," said Charlie. "You forgot the eye shields." Ron took off again.

When he came back for the second time, they began to discuss which of them should inch their way along the crawl space in an effort to get close enough to grab the snake. After due consideration, they concluded that neither of them was dumb enough to try it. So they opened two of the six small doors that opened into the crawl space and each stood by a door. The cobra, attracted to the light coming through the door, picked Charlie's opening, and that was that.

The staff was thrilled to have the snake back. On some days, they had investigated as many as ten calls from people who lived near the zoo, all sure that the snake in their backyard was the one that had escaped. Charlie was called to investigate a dead snake found on Highway 40 that matched the description perfectly. It turned out to be a broken fan belt. He even got to ride in a police car to check out a snake in someone's basement, just behind the furnace. There, he found another broken fan belt. Charlie was always, as he put it, "90 percent sure the snake was in the building." But then, there's always that other nagging 10 percent. Anyway, ten days after the cobra got loose, they decided to reopen the building to visitors. The kids had a ball. Every twenty minutes or so, a child would shout, "There it is!" starting a mad stampede for the doors. Every time it happened, Jerry would cringe.

No one was happier than Jerry when the snake was finally caught. "Talk about a load off your mind," he said. "I lost twenty pounds during that search." Like Steve, Jerry hasn't made a second mistake in over a quarter of a century.

Steve is in charge of the cheetahs, and although they have gotten out twice in the last few years, nobody counts those escapes as Steve's fault. He's still in the "you've had your first mistake" category. Everyone understands that, when sex is involved, strange things happen and, as I said, both times that our cheetah got out, sex was the primary driver.

From studies in zoos, we've concluded that female cheetahs are very selective about mates. This would contrast vividly, for example, with lions, who will mate with whichever male dominates the pride. The Saint Louis Zoo has been very successful pairing males and females over the years, having produced thirty-two cubs in the twenty-five years prior to 2000. Steve wanted to see if Akili, a young seventy-five-pound female, would get along with Chester, a big twelve-year-old male. While they seemed to get along okay at first, in retrospect it seems obvious that the chemistry just wasn't there.

That's why we shouldn't have been so surprised when, on April 15, 2000, a visitor walked up to Katie Kerber, a keeper at the zoo, to pose a question. "Why," she asked, "is it safe to let visitors pet the cheetah?" "I mean," she continued, "couldn't that be a little dangerous?" Katie swallowed once and then asked if the visitor could describe just exactly where she had been petting our cheetah. "Why it's just up next to your bamboo fence!" she replied.

The bamboo fence that blocks our cheetah yard overlook is not there to keep cheetahs in. It's there to keep visitors out. What keeps the cheetah in (at least in theory) is a seventeen-foot-wide moat, about nine feet deep. At the time the exhibit was built, the world record for a cheetah jump was fourteen feet. The designers added another three feet and figured they'd be safe. Low plantings obscure the public's view of the moat, and the forty-two-inch-high bamboo fence is designed to keep visitors from climbing down into the moat and up into the cheetah yard.

No cheetah had ever jumped more than fourteen feet, but as every true sports fan can tell you, stranger things have happened. For example, on October 18, 1968, Bob Beamon leaped into history at the Mexico City Olympics by jumping 29 feet 2½ inches. To put that in perspective, a previous world record, Jesse Owens's 1936 leap of 26 feet 5¼ inches, had stood for twenty-five years, and when that record was finally broken, it was by a mere eight and one-half inches. Yet in one jump, Beamon stretched the world record by almost *two feet.* For the rest of his competitive life, Beamon never jumped over twenty-six feet again—respectable, but still three feet short of his amazing record. So if you wanted to keep a human in an enclosure, you'd take the world record of twenty-nine feet and add an extra three. Thirty-two feet would probably work just fine. But you never know.

Nor will we ever know just exactly how Akili made her improbable leap. Bill Houston, the zoo's assistant general curator, speculates that Chester got a little excited over the prospect of a gorgeous young female in heat (even though they had been together in the exhibit for several days without showing much interest in each other at all) and began to chase Akili. Akili must have been going full speed (approaching sixty miles an hour—a speed that they can reach within only a few strides). She also must have hit the low rockwork near the moat at a perfect angle to make it across. That rockwork has a light, electrically charged wire stretched across it. The wire carries the same charge as the invisible fences that many people (including me) use to keep their dogs in their yards. Before I trained my dogs on our invisible fence, I put the collar on and tried it myself. It was (forgive me) a shocking experience, but it's really not much of a jolt. In Akili's case, she may have hit the hot wire at exactly the right moment to add a couple extra feet to her leap, in much the same way as Bob Beamon did when he took sixteen strides and then hit the take-off board at a perfect angle and at maximum speed.

Finding herself on the other side of the moat was probably as big a surprise to Akili as it was to her keepers. She could have gone over the bamboo fence easily but, finding herself safe from Chester's amorous advances, she simply sat down and rested. That's where she was when Katie found her.

Akili was hand-reared, so it really wasn't much of a trick to get her back into her holding area. The keepers gave her a friendly scratch on the ears, showed her some delicious raw chicken, and Akili followed them back home along the ledge and around to the exhibit's back entrance. Of course, a vet was standing by with a tranquilizer gun, but there was never any real concern that Akili would become aggressive.

Aside from causing us to modify the exhibit moat, there wasn't much public fallout from the episode except for a funny story in the *Webster-Kirkwood Journal*. As luck would have it, our advertising billboards for 2000 featured a cheetah crouched in the grass, with words over the cheetah's head saying, "Keep repeating, 'It's only the zoo, it's only the zoo.'" Beneath the cheetah was the caption, "Live on the edge—The River's Edge." The *Journal* speculated (tongue in cheek) that perhaps we staged the escape to give our advertising campaign that extra "edge." Our marketing director, Kevin Mills, claimed that the timing of the cheetah escape and the ad campaign was, of course, purely coincidental.

The second time a cheetah got out here at the zoo, I was sitting in my office entertaining guests from another zoo. It would be a better story if when the phone rang I had been bragging about how good we were at keeping animals in their enclosures, but I'm sure we were talking about something else. Anyway, in the case of animal escape, my job is to hide under my desk. I've tried it during the periodic drills that we conduct here at the zoo, and although it's a bit tight, I can wedge myself in. Having guests in the office, though, forced me to abandon protocol and feign indifference. "Gee," I casually said, "it's been three years since that last happened, but I'm sure we'll get the situation under control. I'd go out and help, but as a good host, I should probably stay right here with you guys . . ." (I'm joking, of course. I have a job in the event of an animal escape, but it's not very glamorous. I work with the PR folks to try to ensure that we get accurate and timely information out to the media. You can get hurt doing this, too, but usually it's no worse than a paper cut.)

Again, the paper had a field day. They reported that the incident resulted from an arranged date between Halala, a three-year-old female, and Shanto, a two-year-old male that was just barely sexually mature. The staff introduced the two animals and watched them quite closely for an hour. Shanto, according to the keepers watching the introduction, seemed to confuse flirting and playing. Whatever his intent, he backed Halala into a corner at a spot the observers couldn't see, and she presumably used some thistle and other vegetation to help herself over a ten-foot wall. Since she was out of sight to begin with, the keepers never knew she was out of the exhibit—but some rather startled visitors noticed. "It was quite strange," said Jan Ikemeier, who was visiting the zoo with her two daughters. They were on their way to the bathroom when they heard some rustling in the bushes. "The man next to me said, 'Oh! It's just a bird,'" she recalls, mimicking his dismissive tone. "I came over a bit and looked, and saw these legs . . ." The cheetah was "very calm," Ms. Ikemeier noted, saying, "I would have petted her if I could, but instead I told the girls to back up slowly. Once I got out of sight, I ran like crazy until I could find a zookeeper."

Ms. Ikemeier did everything right. About the only thing that could draw a cheetah to attack would be to see small prey (like children, for example) running away. At that point, a cheetah's instinct could kick in with some bad results.

At Indianapolis we had a tiger escape not once but twice. Tigers are, in comparison with cheetahs, a much different matter. An adult tiger is extraordinarily dangerous, and all zoos have a similar policy for dealing with the escape of a dangerous predator—they kill them.

Every zoo that I know of has teams of "shooters" who are trained to respond to animal escapes. There are guns in strategic places around every zoo campus. Beyond that, there is considerable variation in policies. For example, at the Saint Louis Zoo, our policy is that, if a dangerous predator should escape, the police are called—however, the zoo staff calls the shots (so to speak). The police are not in charge. At Indianapolis it was (and still is) different. The Indianapolis Zoo staff trained with the Indianapolis SWAT team on a regular basis, using the police range to practice with their SWAT counterparts. If an animal was to escape, they would go out in teams of two, with a zoo staff member paired with a SWAT member.

The second time we had a tiger escape, though, the police were never called. It was five years after I arrived in Indianapolis, and our female Siberian tiger, Lena, had recently delivered her first set of cubs. They were old enough to play out in the exhibit, and when they weren't sleeping with mom, they were little bundles of manic energy. They climbed, leapt, chased, and chewed. It was the chewing that got us into trouble.

At Indianapolis, the tiger exhibit had piano wire stretched across the viewing areas, where the public stood, and tall rock walls designed to obscure the tiger holding area behind the scenes. Since tigers are great climbers, the designers had installed wire mesh above and over the rock walls. They guaranteed us that an adult tiger couldn't get his or her teeth into the wire mesh and open a hole large enough to escape. They did not consider, though, that a baby tiger might be able to.

Sure enough, a few weeks after the cubs began to explore their outside habitat, one of the little guys gnawed through the metal mesh and crawled out onto a ledge above the public viewing area. From there, he was able to enjoy the sight of the visitors walking just below. Fortunately, a visitor saw him, called a keeper, and she scooped the baby up and returned him to the holding area to be reunited with his siblings.

Of course, a baby tiger poses no real danger (although you might want to wear gloves when you pick one up—we've already seen how strong those little jaws are), but an adult is a different matter.

What's interesting about the escape of the baby is that it cleared up a mystery that had plagued the staff for quite some time: How had the first tiger escape come about?

The story begins on Sunday, February 3, 1991, at 9:20 PM. That is when Lew Birmelin, who organizes the Ralph Ned teams at Indianapolis, got a call from security saying that a Siberian tiger was seen running loose on the grounds. The tiger, a big male named Chudal, had been on the rocks of his exhibit for the previous two nights and showed no interest whatsoever in coming down, even preferring to skip meals in favor of his lofty perch. Now he was running loose.

When Lew and his team had arrived, at about 9:40, he was informed that one of the security officers had the tiger under observation (from the safe confines of his Jeep). Allen Monroe and Keith Schnell, both on the zoo staff, joined Lew, and all three of them made their way toward Chudal, who was crouching about fifty yards west of his exhibit.

No one on the staff would ever want to harm an animal, but trying to dart a cat that big and dangerous, especially one that is out of its primary containment and could easily scale the perimeter fence and be out of the zoo entirely, would place the staff and the general public at an unacceptably high risk. So Allen was using slugs. He had the shot, with the tiger only 150 feet away, and fired three times.

He missed with all three shots (or so it seemed at the time). The tiger, apparently startled by the noise, raced along the perimeter fence toward the back door of the zoo's gift shop, located some fifty yards east of his exhibit. Shortly after that, he turned back toward his exhibit and disappeared into the deep bamboo that lined the public walkways.

The staff watched anxiously into the night for the tiger but could catch only fleeting glimpses from time to time. By 10:40 in the evening, they still had not gotten close enough to try another shot. At this point, Debbie Olson suggested driving up to the back of the exhibit and opening the gate to the exhibit yard. Her logic was pretty simple—that was the tiger's home and he would likely be looking for a safe place, given all the noise and excitement. Debbie drove up to the gate with a member of the SWAT team, who reached out and opened the gate as Debbie slowly backed up until the gate swung wide.

It wasn't until 4:30 in the morning that our tiger decided, in effect, that there's no place like home. He suddenly appeared at the entrance of his exhibit and then glided noiselessly through the open gate, all the while under the watchful gaze of the staff. At 5:10 he was seen moving into his indoor holding area, so Debbie pulled up with her vehicle and pushed the gate closed. Dave Merritt and Lew then went in the holding area and closed the door that led out to the exhibit (while Debbie secured the gate), and our tiger, finally, was back in safety.

That day the vet staff went into the holding area to see how Chudal was doing. They anesthetized him and began the process of carefully examining the now-sleeping animal. They were shocked to find that one of the three slugs from the 12-gauge shotgun had hit Chudal squarely on his scapula. The slug was still embedded under the tiger's skin, but it had not even broken the bone. They removed the slug and stitched the wound. That was when Dave Merritt, now the general curator at the zoo but then the curator of marine mammals, felt fear for the first time: "We were out there thinking that at least we were safe. We were armed, after all, and thought, if worst comes to worst, we can always shoot. But when I realized that even a jacketed slug from a big shotgun had no real discernable effect on Chudal—that's when I realized that I was never really safe—not the whole time."

Apart from Dave's dawning awareness of his own mortality, the zoo staff never learned how Chudal made a hole in tiger-proof mesh, not until years later when they discovered that he hadn't. His cubs had started the hole for him. Once he realized that there was a place where he could get his teeth around the mesh, he opened the rest of the hole for himself. That is what had kept him on the rocks for two days and away from his dinner. He had a little project going up on those rocks and wanted to finish it. Thankfully, he survived his adventure pretty much unscathed.

Earlier I said that my job, in the event of an escape, is to hide under my desk—well, work with the media and with PR staff—but I did get close to the action in the recovery of one escaped animal not long ago. On October 9, 2002, our PR director walked into my office with the unwelcome news that a muntjac had left the zoo grounds and that camera crews were covering the big breakout.

A muntjac is a tiny member of the deer family that ranges from the forests of India and Nepal all the way to the Malay Peninsula. It's about the size of a beagle, weighing between forty and sixty pounds and standing about two feet high at the shoulder. They are deep brown in color, have small antlers, and the upper canines of the males are elongated—almost like tusks. Both males and females have a surprisingly deep alarm call that sounds like a bark and carries quite a distance.

We first discovered that one of our males had left the zoo when an early morning visitor knocked on the door of the camel barn to tell one of our keepers, Chet Stuemke, that one of our little deer was standing outside the zoo on the corner of Wells Drive looking a little confused. That part of the zoo perimeter fence is obscured by a thick stand of bamboo, and as soon as the animal staff arrived on the scene, the muntjac disappeared into the safety of the dense plantings.

That allowed the staff plenty of time to arrange themselves around the stand of bamboo, using baffle boards and a cloth fence made of burlap to keep the animal trapped. It also gave our vet, Luis Padilla, time to prepare a mild tranquilizer. Finally, it gave the media plenty of time to show up for a live shot. (My job was to stand around and be ready for media comment.)

Luis actually darted him twice, but either the dart didn't inject the sedative, or it just didn't stay in long enough. In any case, after about a half hour, Martha Fischer, our curator of ungulates, snuck into the bamboo, hid herself along a little path, and grabbed him as he walked by. That's when I learned that muntjacs can bark so loud. Martha put him back into his stall in the barn and then began the process of figuring out how he had gotten out on the street in the first place.

The first thing the staff did was to walk the entire perimeter of the exhibit yard, looking for breaks or gaps in the fencing. They found nothing. Next, Martha sat Chet down and went through every move he had made that morning. Chet said he clearly remembered securing every gate and shift door and, in fact, when they went in to check, the gates and shift doors all were closed and pinned.

Having eliminated the possibility of a perimeter break and keeper error, they were faced with three remaining possibilities: the muntjac could have jumped the perimeter fence, or jumped the front wall of the exhibit and gone through the visitor turnstile exit, or maneuvered under four different gates and walked out the back door of the camel barn and out to the street. But none of those possibilities really made much sense. There was a three-inch gap between the four gates and the ground, but muntjac have four-inch antlers. The perimeter fence is chain-link, eight feet tall, and has bamboo growing behind it that makes it appear taller still. That would have to be, if nothing else, an imposing visual barrier.

The best explanation could be that he leaped out over the front wall of the exhibit, which would mean a jump of about seven feet, or roughly four times the height of a muntjac. It's theoretically possible, but he had been in this huge yard for years and was never a particularly excitable animal. Plus, even if he cleared the moat wall, he still would have had to go through a full-size exit turnstile to get off the grounds. It's clear that we did everything right in terms of our exhibit and daily protocols, and the staff did a brilliant job getting the muntjac back, but we still don't know exactly what happened. Just like Indianapolis's first tiger escape, our muntjac's escape might have to remain a mystery, at least for a while.

Actually, we have one species that leaves the zoo grounds on a fairly regular basis. We've always had peacocks wandering the grounds, and believe it or not, peacocks are pretty good fliers. Well, that's something of an overstatement, but they *can* fly. They fly considerably better than, say, a chicken.

Ours go out of the zoo once or twice a week to explore Forest Park and, if we have time, someone goes out to round them up and shoo them back through one of our perimeter gates. The problem with the loose peacocks is that they give the park police all kinds of fits.

A peacock's call sounds amazingly like a woman screaming frantically: "Help! Help! Help!" Granted—it sounds like a very large woman, with a fairly deep voice—but it sounds pretty convincing all the same. I don't know how many times the park police have been called to investigate this kind of false alarm, but I do know that they tell all the rookie park policemen that if they ever hear a woman screaming for help near the zoo, the odds are astronomically high that it's not a woman at all.

My favorite story of an animal escape, at least at our zoo, occurred when one of our chimps, named Kumi, escaped on May 2, 1987, which just happened to be Meg White's first wedding anniversary. Meg still works with our chimps, and she was around when our current ape house, Jungle of the Apes, was first opened in 1986. Shortly after the building was opened, we realized that none of the pins in the hinges of the holding doors had ever been welded down. This meant that a clever gorilla, orangutan, or chimp could conceivably take the pins out of the hinges and walk out, a free primate.

Of course, we immediately called in a welder, and he set to work fixing the problem. Naturally, he had to undo all of the doors and the process was a huge disruption for the staff. There are twenty-four doors with three hinges per door, so the whole effort was also fairly time-consuming.

Kumi, who was about ten years old at the time, was born at the Saint Louis Zoo, then was moved to Lincoln, Nebraska, and then back to St. Louis. She had been hand-reared and was socialized to humans, so we were in the midst of reintegrating her into our chimp social group. It was a long reintroduction, and Kumi couldn't go out in the exhibit with all of the other chimps at the same time, so on May 2, she was the only animal back in the holding area. Our welder went off to lunch, and Meg, feeling a little sorry for Kumi, opened all of the shift doors so that the chimp could have the run of the entire interior holding space.

The problem was that the holding space, consisting of six individual caged areas connected to one another by the shift doors, is in the shape of an L. Meg could see that the four doors that opened out to the keeper area were all securely closed, but what she couldn't see was that the welder had left one of the doors to the keeper area, on the other side of the L, wide open. Kumi, on the other hand, noticed immediately.

When Meg came around the corner, the first thing she saw was Kumi, standing outside the cage some ten feet away. "Hi, Kumi," she said in her most pleasant voice. Then she started to back up toward the gate that led out to an empty exhibit yard, reached behind herself for the latch, and let herself out of the building, leaving Kumi alone inside.

Our other keeper, Mike Lynch, was working in the office of the ape area, which is separated from the keeper area by large security gates. Most of the time, the staff leaves the gates ajar because it's a giant pain to constantly open and close them every time they want to go to the office. Meg, now standing out in the yard, called out (again in her sweetest, most nonthreatening voice), "Oh, Mike! Kumi's out. You might want to shut the security gate!"

Thankfully, Mike heard her, closed the gate, and then called the emergency number. He was soon joined by our vet, Randy Junge, our curator Roger Birkel, and several others. The first thing they did was open the exhibit gate so that Meg could get out of the yard. Then Roger sidled up to the security gate armed with a syringe of ketamine, a mild tranquilizer. Roger knew Kumi pretty well and Kumi, socialized to humans, kind of liked Roger, too. Roger called her over and, when Kumi reached through the bars to give Roger a friendly hello pat, Roger grabbed her hand and tried to inject the tranquilizer. He missed. Again he called Kumi, but this time she wasn't nearly as willing to come up to him. After all, she'd just gotten stuck with a needle and wasn't too keen on repeating the procedure. Roger did get her over and, again, grabbed her hand, but having been through this once before, Kumi was off like a shot. Plan A had failed miserably. Kumi obviously was not coming back for a third try.

So the staff went to Plan B. They ordered pizza from Imo's. If you've never had an Imo's pizza, you really should try one. They're made here in St. Louis. They have an incredibly thin crust, and they're topped not with mozzarella, but with provel. They're an acquired taste, but once you get used to them no other pizza will do.

The pizza party was part of a broader strategy, though. By holding the party just outside the security gates, they figured Kumi would naturally come over at some point to mooch a little snack. In so doing, she would have to pass by Dr. Junge, who had taken Roger's place in the gorilla tunnel, this time with a blowgun instead of a handheld syringe.

Kumi, of course, was interested in the pizza, but she had other things she needed to do before she joined the party already in progress. For a year,

Kumi had been watching the staff working in the keeper area, and now that she was in the area for the first time, she had some things she wanted to try.

The first thing was talking on the telephone. She reached up, took the receiver off the hook, and held it to her ear. Hearing nothing but a dial tone, she tucked the phone up to her ear with her shoulder and went back to work. That actually pleases me in a funny way. It tells me that our keepers continue to work, even when they're talking on the phone.

Her next job was to clean the drains. Our building has small round grates over every drain, and under the grate is a mesh basket designed to stop the bigger stuff from getting into the sewer lines and clogging them up. Kumi had no problem getting the grates up and the mesh baskets out. She didn't realize, however, that the phone cord would only stretch so far, so the phone was, in a literal sense, disconnected.

Next, she went into the storeroom to get out the hip boots that the staff wore to hose the exhibits. She tried them on and apparently decided that they were either uncomfortable or unstylish, so she took them off. Of course, hip boots weren't the only thing in the storeroom, and Kumi soon discovered that light bulbs make a terrific popping sound if you throw them hard enough.

Kumi actually found plenty of other things to do, so it wasn't until about midnight that she came over for pizza. By this time, Randy was getting a little tired of holding his blowgun, but as she walked by, he had one chance to dart her and he did it.

Moving on, Kumi disappeared out of everyone's sight; after about ten minutes of silence, Roger decided that he would go in and check on her. Sure enough, she was already fast asleep. The good news was that they never did have to share any of the pizza.

The most fascinating ape escape story, though, involves Fu Manchu, the orangutan that we met earlier. Fu's escape story is legendary, but the best-written treatment I've ever seen of it is in Eugene Linden's book, *The Octopus and the Orangutan*. Linden's book tells a number of stories about animal intrigue, intelligence, and ingenuity, but he leads with Fu's story for a good reason—it is as baffling as it is wonderful.

In a nutshell, what Fu did was to pick the lock on his cage door, not once, but multiple times, by sliding a bent wire in between the door and the jamb, rotating it to push the tumbler back, and throwing open the door. He hid the wire in his mouth between escapes, and it drove the keepers crazy trying to figure the whole thing out.

As Linden pursued the story, it got even more bizarre. Clearly, Fu had demonstrated three pretty incredible traits. First, he had made a tool, in the form of a bent wire lock pick. Second, he had successfully used the tool—not once, but multiple times. Third, he had used an intellectual ability that most of us don't realize that other primates have—he had used *deception.* He had successfully hidden the wire in the one place where the keepers were singularly unlikely to find it—his mouth. But that isn't the end of the story. We'll never know for sure, but Linden thinks, after talking extensively to the folks at the Henry Doorly Zoo, that Fu may have gotten his tool from the orangutan next door, a lovely but very overweight female named, ahem, Heavy Lamar.

The staff discovered that the wire Fu used was the same gauge as the wire in a damaged light fixture in Heavy's cage. Heavy was being kept apart from the other orangutans because she was, well, heavy. They had her on a strict diet, and the only way the staff could keep her from wolfing down everyone else's food was to keep her apart from the troop until she could slim down a bit.

Before the series of escapes began, the keepers recall Fu slipping Heavy some of his monkey biscuits through the wire mesh that separated her from the other orangutans. Linden offers us several different possibilities to consider, now that we know that Fu was supporting Heavy's monkey biscuit habit. First, it is possible that Heavy begged Fu for a biscuit, Fu pointed to a piece of wire that Heavy had stripped from the broken fixture, and they made an even-up trade, with Fu having no clear idea of what he would do with the wire once he got it. Second, it is possible that Fu had already figured out exactly what he would do with the wire, and thus set about bribing Heavy, his escape plan already in mind. Third, and admittedly unlikely, Fu figured out how to bribe Heavy to vandalize the light fixture, communicated his need to his chubby friend, and paid her off with the one thing she couldn't resist—food.

We'll never know for sure, but there are literally hundreds of examples of trading between orangutans and their human caregivers. Keepers trade with orangutans all the time for things they shouldn't have, so why wouldn't orangutans trade with each other? Of course, it's possible that Heavy just stripped the wire on her own and accidentally dropped it in Fu's cage, and that Fu was slipping her monkey biscuits out of the goodness of his heart. My friend, Dr. Lee Simmons, who runs the Henry Doorly Zoo in Omaha and knew Fu well, will tell you this much—he's not sure what happened, but he does believe that Fu was a pretty magnanimous orangutan.

～

I can't resist one more escape story, but this one is considerably shorter. There is a place called Wildlife Safari located in the town of Winston, Oregon, which is about 160 miles as the crow flies from the next closest zoo, in Portland, Oregon. One morning the phone rang at Wildlife Safari. It was the police calling to inform the staff that they had found a hippo munching grass in the middle of the town. John Gobershock, a manager at Wildlife Safari, took the call, and his response will live on forever in the annals of zoodom: "Are you sure that it's *our* hippo?"

10

Immaculate Conception

I'm sure everyone has a favorite example of political correctness gone awry. My first experience with the PC movement dates back to the early 1970s, when I was in college at the University of Missouri. Our student newspaper was named *The Maneater*, after Missouri's mascot, a tiger. A vocal (and, thankfully, small) segment of the student body wanted to change it to *The Person-Eater*.

The PC movement struck the elephant research program at the Indianapolis Zoo much later. Debbie Olson, the Indianapolis Zoo's curator in charge of elephants, decided, in the late 1990s, that I ought to give serious consideration to dropping the phrase *artificial insemination,* and going with *assisted reproduction,* as it was more politically acceptable. I wasn't sure if she was serious, and my face must have reflected my confusion. "Look," she said, "it puts all the emphasis on the semen, and besides, there's nothing artificial about it." I suppose I agree with her, but old habits die hard. I still use the old terminology, although I'm gradually coming around to Debbie's point of view.

The Indianapolis Zoo started their assisted reproduction program when it was still called artificial insemination. In the late 1980s, Debbie, along with some of the more progressive trustees of the zoo, began to map out a strategy for research. The program was not on the front burner, even though the group of advisors who helped to plan the zoo thought research was integral

162

to any modern facility. All of the attention was focused on getting a new zoo built and open. Research simply had to wait.

When I arrived in 1993, the staff made it quite clear that they had waited long enough. The new zoo had been open almost five years, and although they still had many problems (most of them budgetary), the potential for doing some vitally important work was clearly there. Debbie was normally very quiet and restrained, but when she got on the subject of research, she became pretty emphatic. Not only did she think it was time to go to work on our research program, she was also quite sure that our elephants should be the focus of our initial efforts. The problems were, first, that there was so much we didn't know about elephants that it was hard to know where to begin, and second, that there was so much that we were uniquely positioned to do.

In the early 1990s a number of field studies had been done on elephants. Iain Douglas-Hamilton, along with many other very talented research scientists, had spent years following elephants in the wild, documenting their behavior. Field researchers learned an enormous amount about elephants just by watching them (although there are still many mysteries left to probe), but elephant biology in general, and reproductive physiology in particular, was still poorly understood. In places where the population of elephants was strictly regulated, such as the Kruger National Park in South Africa, elephants were killed when their numbers got too great. Vets came along behind the hunters and did necropsies, and we learned quite a bit about fetal development and gross physiology from their efforts, but we still didn't know much about how the bodies of living elephants functioned.

I'll talk later on about killing elephants in order to maintain their numbers, but while I'm on the subject, I should say that this occurs much less often now than it did, say, ten years ago. There are clear signs, however, that culling will increase again in the near-term future. It is now possible to relocate elephants, so when their numbers get too large in one place, they can be moved to another locale where there are fewer elephants present. This is not a particularly good solution to population control, however, as it is costly, the number of animals that can be successfully relocated is small, and, unless they are moved a very great distance, they often simply return to their home range.

In most of the places in Africa where elephants were historically found, they have been hunted to extinction. These days, the increase in ecotourism in Africa, coupled with an increase in the number of private and public preserves, has meant that there are now a few more places for surplus elephants. There simply weren't as many reserves ten years ago; even if there had been as many, we didn't know how to safely and inexpensively move such large animals. The technology really wasn't perfected until 1992, when a drought in Zimbabwe forced the conservation community to try to relocate adult elephants from Ghonarezhou Game Reserve in southeastern Zimbabwe, where the drought was most severe, to parks in South Africa, as a last-ditch effort to save them. The number of reserves has not increased enough, though, to make translocation a real solution.

In order to get started with research on elephants, Debbie convinced the administration, shortly after the zoo opened, to allow her to offer elephant rides to the public *and* to keep 10 percent of the revenues from the rides to fund elephant research. (The administration had originally offered the staff a 10 percent raise; the staff declined the raise in preference to a research budget.) The rides generated about $100,000 per year, and her $10,000 went into a special account that built up pretty fast. The result of Indianapolis's ride program was threefold: the research program got seed money, the staff became very enthusiastic about giving rides, and our elephants got a lot of exercise.

At the same time, Debbie continued talking to other zoos, field researchers, and conservation officials about issues related to elephant research. As a result, she made a huge number of new national and international connections, and she enhanced existing connections, within the community of people who were interested in and supportive of the future of elephants in Africa, Asia, and in America. The more she talked to people around the world about elephants, the more she began to understand what the real research needs might be.

Her conversations led her to conclude that there were some serious issues that needed to be studied. First, it became clear that we didn't know much about health and disease in elephants. In the wild, elephants aren't really a prey species unless they are very young. They can live a fairly long life if they survive their infancy and aren't killed by poachers. The popular notion that their life spans are only slightly shorter than humans' is, however, largely a myth. According to Cynthia Moss, a well-known elephant researcher, the average life expectancy in the wild at birth is about thirty-five years for females and somewhere in the early twenties for males. It is true that they can live fifty to sixty years in the wild, but then it's true that humans can live ninety to a hundred years, even though few of us do. They don't really die of old age, though. Just as we humans have two sets of teeth, elephants have six.

Molars slide down their jaw, two on each side, top and bottom, and replace one another as they are worn down by constantly chewing the tough grasses. When the last set is worn to a nub, they can no longer eat, and they die. In fact, the popular myth of the elephant graveyard may have some basis in reality. Old elephants whose last set of teeth is worn down will move to a swampy, low area where the vegetation is softer. Eventually, though, they can no longer chew even the softest vegetation. But elephants, like any living creature, are also subject to disease, and their bodies change in other ways as they mature and then grow old.

In zoos, foot care is a major issue. As I mentioned earlier, elephants walk on their tiptoes. If you make a tepee with your fingers on a table, you'll get the idea. The area within the radius of your fingertips is the equivalent of the bottom of an elephant's foot. In elephants, this area is filled with a kind of spongy cartilage. In the wild this cartilage is constantly worn down because the animals are forced to walk so far to find food. In zoos, where food is close at hand, the bottoms of their feet do not wear as fast, so their toenails and soles have to be trimmed to maintain good foot health. Foot health is critical for an animal that can easily weigh five or more tons.

So health issues were one area that needed study. A second need that emerged centered around a desire to standardize training and husbandry techniques for animals in zoos. This was important because it would otherwise be difficult to move animals from one zoo to another. Moving animals, in turn, is important from a genetic point of view. With a relatively small population of animals, it is important to maintain gene flow. In the absence of gene flow, you get inbreeding, which creates problems.

The third need she documented was, perhaps, the most acute. In the early 1990s there simply weren't enough African elephant babies being born to sustain the population of elephants in America. Demographers often talk about something called "the age pyramid." In a growing population, the base of the pyramid (the number of infants and young) will be very big. At the top of the pyramid, the number of very old individuals will be very small. The shorter or squatter the pyramid, the faster the population size will grow. The narrower or taller the pyramid, the slower the population size will grow. What Debbie documented was that the base of the pyramid for elephants in the United States and Canada was almost nonexistent. In other words, if you graphed the population of elephants by age, it would actually look like a pyramid turned upside down. In the fifteen years between 1978 and 1993, only twenty African elephant babies were born in North America. Having almost no babies meant that the population, from a demographer's point of view, was going to die out. We needed to make more.

In the old days, zoos would not have worried so much about that. Twenty or thirty years ago we just would have said, in effect, "So what? We'll just go to Africa and get more." Today, with our emphasis on keeping genetically viable, stable populations in zoos, we are much less willing to extract animals from the wild, even if they are considered to be surplus—that is, too numerous for the local habitat to support.

I mentioned that elephants are culled in many areas of Africa. *Slaughtered* would be a better word, but as much as I detest killing animals, I also have to admit that, if there are too many elephants in too small a space, something has to give. There is a legitimate debate concerning the role of elephants in maintaining a natural ecosystem. Some biologists argue that, in their natural state, there could never be too many elephants. The argument goes something like this. Sure, elephants cause an incredible amount of change to an ecosystem. They move through an area, stripping the delicious bark off of living trees, which slowly kills the tree, or pushing smaller trees down so they can more easily eat the leaves, which quickly kills the tree. A large herd of elephants can destroy a wooded area in a very short time. If, however, they were really the agents of wholesale ecological destruction, they would have destroyed most of Africa centuries ago, so it follows that they must help the ecosystem at least as much as they harm it.

Nature, if left to its own devices, will always find a balance. The problem is that nowhere in Africa is nature left to its own devices. As humans have spread throughout the continent, Africa has been completely transformed. If you don't believe me, fly over Kenya sometime. The outlines of the national parks are perfectly visible from the air. The only wild places left are those that are protected from human incursion, and those places are really very small. If the Africa of today was anything like the Africa of two hundred years ago, we could have a very interesting debate about the role of elephants in shaping or maintaining the ecology. But it isn't, so we can't. Elephants are confined to relatively small, isolated parks and preserves. If their numbers grow too large, the effect of their feeding patterns will be quickly felt. Not only will their influence be seen on the vegetation in their preserve, but also they will be forced to look for food outside their preserve. The agricultural fields that abut a typical Kenyan preserve on every side become enticing targets for the herds. As soon as they begin to browse outside the park, there are, not surprisingly, demands to reduce the size of the herd. The bottom line is that, even if we didn't need to cull herds in order to protect the vegetation within a park, we would soon have to cull herds in order to protect the agricultural areas and human life outside the park.

Even if we had a ready supply of infants available to us from the wild, this still flies in the face of what we're trying to accomplish in zoos. We're supposed to be safe repositories for species that may, or may not, survive in the wild. If we can't do that, then we are not doing our job. On the other hand, if we can keep a genetically viable pool of animals, we could at least maintain the species and, perhaps, someday reintroduce it back to its native habitat. Even if the species is not extinct in the wild, we can help facilitate gene flow by insuring that animals move back and forth between small wild populations and the populations in zoos.

It seems clear, though, that given the existing demographic pyramid in North America, we will have to import at least some animals from the wild. More importantly, we will need to dramatically increase breeding in zoos.

Of course, there are other reasons to keep animals in zoos apart from the need for a stable, genetically viable safety-net population. Education is a critical one. People don't care about abstract issues as much as they care about things that they have seen or felt for themselves. Putting them in front of a real elephant makes an abstract problem a real issue. Research is another reason. There is so much we don't know about elephants. We don't know how good their memory is (although it is popularly believed to be pretty good as witnessed by the common "you have the memory of an elephant" figure of speech). We don't know how well they can see. Some people suspect that they have what we might call a primitive sonar system, because they seem to know precisely where to dig for water in order to find it within the range of their long trunks. We know very little about their communication system. The list goes on and on. Some things, like the discovery that elephants could communicate with one another using a frequency so low that humans cannot hear it, were discovered in zoos. In that case, a researcher felt a resonance in the chest and reasoned that it might be a sympathetic vibration from an elephant call. Elephants were then taped "speaking" at their low frequency and the tape was speeded up in order to make their calls audible to the human ear. Now field researchers are taping the calls of elephants in the wild and beginning to unravel their speech.

By the time I arrived in Indianapolis, elephant specialists in zoos around the nation were working toward standardizing training techniques, and a good deal of information on how best to care for elephants had already been gathered. However, very little work had been done on the central problems associated with elephant reproduction. Debbie and the staff of the Indianapolis Zoo had spent the better part of five years training our elephants, and they had developed a remarkably high level of trust in our five elephants, and the elephants, in turn, trusted their keepers. The elephants had learned to stand comparatively still while people worked in close proximity. (This is what saved them when they contracted salmonella. The keepers could stand next to them day and night, holding bags of glucose and saline solution on long poles while the rehydrating mixture flowed slowly into their bodies. This and this alone kept the elephants alive, but they never could have been treated if they weren't so trusting of the keepers.)

It was clear that we could work closely with the elephants, and it was also clear that we had to solve the riddle of getting more babies on the ground if we were to keep a self-sustaining population in North America. The reasons why the elephant population is small in North America are complex. In the first place, there aren't many bulls, and several of the bulls we do have are not reproductively viable. The reason why we have so few bulls is that they are very hard to keep in zoos. In part, this is because of their size—upwards of fourteen feet tall at the shoulder, and seven to perhaps eight tons in weight. More importantly, bulls have a breeding season, or *musth,* when their testosterone levels shoot up dramatically. During this period they go, well, a little crazy. This, in combination with their great size, makes them very difficult to work with.

In the wild, bulls live a fairly solitary life. As teenagers, they are evicted from the herd of closely related females and move off to graze on their own, forming loose associations of bulls of their own approximate age. They will occasionally pair up with one or two other bulls, but these temporary partnerships end when one of the bulls goes into *musth.* This is the only time that bulls usually get to breed; being in *musth* gives them the wherewithal to keep the other bulls away from any cow in estrus. The high levels of testosterone give them incredible strength and endurance and, most importantly, rage. I have watched two bulls in *musth* fight all day long. They eye each other warily and then display their enormous strength by pushing over a tree and stomping the branches to kindling. They bellow at each other and then charge. They grapple with their trunks and smash their massive heads together and then part again. They hurl dirt in the air and stalk each other relentlessly. Eventually the dominant bull drives off all challengers and mates with the cow.

In zoos it is very hard to keep a bull when it is in *musth*. They will charge anything in the wild and they behave in precisely the same way in zoos. They hurl whatever they can get their trunks on, smash into walls, and destroy anything they can. They may be charming and gentle when they are not in *musth,* but when they are in *musth* they are, quite simply, not responsible for their own actions.

As any biologist will tell you, it is still possible to have plenty of babies even if there are only a few males. All we would have to do is move the females to where the bull is, right? Well, it turns out to be much more complicated than that. For one thing, it is just plain difficult to move an elephant. More importantly, it is stressful to separate a female from her herd if she is not used to traveling. Circus elephants, who change locations regularly and are used to traveling, cycle normally. But in the case of elephants raised in zoos, a female often will stop cycling as a result of the anxiety associated with leaving her family. So, to mate a female, we have to move her to a new home, get her adjusted to her new surroundings and new herd, wait for her to cycle, and then wait for two years (actually, twenty-two months) for the female to gestate and give birth *or* move the cow back before her third trimester. The upshot of all this is that it takes at least a year or two to send an elephant off to be bred and return her home. It can be done, but it is clearly very difficult.

On the other hand, it would be relatively simple to just move the bull to where the cows are, but this entails a problem, too. The bulls are huge and require special facilities. They must have their own holding space, and it must be substantially stronger than the holding area for cows. They must have their own separate exercise yards, and if they're going to be on display for the public, they require a separate exhibit. Most zoos simply don't have the space or the budget to build separate facilities for bulls that might only be on-site for a limited amount of time.

The obvious solution to all of this is to just move the semen. If we can get the bull to donate semen, we can artificially inseminate the cow. The cow doesn't have to go anywhere, and the bull can stay at home as well. This sounds simple, but again there are problems.

In the first place, elephants are what we call "silent ovulators." They give no external sign whatsoever that they are ovulating, so it is practically impossible to know when to inseminate them. Obviously, the bulls know, but if you don't have a bull around, you just can't tell when the time is right. In the second place, it is very difficult to get the semen through the hymen to the cervix. (In elephants, the hymen is not broken when the cow has sex for the first time. It is broken when the baby is delivered.) In cattle, it is a relatively straight path from the entrance of the vagina to the cervix, and the distance is only about as long as your arm. In elephants, the distance is more like six feet and, worse yet, the path is shaped more like an S. You have to first go up over the pelvic rim, take about a 90-degree turn downward, and then up again. Third, even if you could find the hymen, you would find three holes in it that appear to lead to the cervix instead of one. Two of the holes are blind. They don't go anywhere. So you have to find the correct hole in which to deliver the semen. Finally, it is impossible to "look" into an elephant using conventional ultrasound. We can't see where we are as we insert a semen-bearing tube inside the animals from outside their bodies (as we do with ultrasonography in humans) because the bodies of elephants are so big and their hide is so thick.

Debbie's first problem was to figure out when the cows ovulated. She set out to solve this problem by taking blood from the ears of the cows and analyzing the blood for the presence of the hormone progesterone. By taking blood every week from each of the cows, she determined that the elephants had an estrous cycle that peaked about three times a year. Still, that didn't tell her precisely when to inseminate. In order to understand when to deliver the semen, she had to determine the precise time that the egg was released by the cow. To determine this, she started collecting daily blood samples and analyzing them for the presence of luteinizing hormone (LH)—known as the "egg-producing" hormone. This project was done in conjunction with Professor Malve at Purdue University. Eventually she checked for LH multiple times during the day, to look for evidence that circadian (twenty-four-hour) rhythms might be affecting the elephants' LH. Due to this extensive testing, she found that, within the estrous cycle, elephants produce shots of luteinizing hormone twice. The first peak occurs about twenty days before they actually cycle, and the second peak, twenty days later, actually causes an egg to be released, making the female receptive to breeding.

At this point, we are unaware of any other mammal that has two LH peaks. No one understands why elephants do this, and this particular discovery is making field researchers take another look at the data they have been collecting over the years. It could be that the first peak is a signal to the bulls that they should move toward the herd, as there will soon be a female ready to breed. Another theory is that perhaps the younger bulls, which have not yet been evicted from the herd, are exposed to the scent of fertility, and perhaps they even practice on the females even though nothing will, biologically speaking, come of their efforts.

From the zookeeper's point of view, the significance of the first peak is that we have almost three weeks' warning that a female is about to produce an egg. That gives us plenty of time to line up a bull to donate sperm, make plane reservations so that we can fly the sperm from the donor's zoo to our own, and line up all the assistance for the actual insemination procedure.

It took five years of painstaking data collection and analysis in order to document the estrous cycle and the LH cycle. While we were collecting blood for analysis, we were also training the animals to undergo the actual insemination procedure. The techniques used were developed in East Berlin by the scientists at the Institut für Zoo und Wildtierforschung (IZW), or, in English, the Institute of Zoo and Wildlife Research. The German scientist who was primarily responsible for figuring out how to get the semen to the cervix was Dr. Thomas Hildebrandt. Thomas is a brilliant scientist, but he was not a good Communist. Growing up in East Berlin, he wanted to be a physician, and because he was so smart, he felt that he could make his political feelings known and not suffer any repercussions. He was, as it turned out, wrong. He was exiled from medical school and took up work at the zoo in East Berlin. Eventually he was reinstated in school, but by that time he had become more interested in animals than in people. After he finally got his degree, he went to work at IZW and, along with Dr. Frank Goeritz, he developed a new technique for using ultrasound on elephants that got around the problem of thick hides and dense bodies. His solution was to ultrasound the elephants by reaching a specially designed probe far past their anus. This allowed the researchers to, in effect, look down from inside the elephant and guide a tube directly to the cervix. Thomas worked on developing a special ultrasound probe extender, which he used in conjunction with a vaginal endoscope, used to deposit semen in the correct location. The equipment was manufactured to his specifications by a company with the improbable name of Arno Schnorrenberg Chirugiemechanick along with Ruesch AG, both located in Germany, and patented in 1995. All he needed to make his technique work was an elephant that would stand still for this procedure. This, in turn, required some very well-trained elephants.

For two years Debbie and her keepers conditioned their cows for this procedure. They started by asking the cow to hold still while they gently touched her around her anus. Gradually, the elephants got used to this attention, and the keepers moved on to inserting their (gloved) hands up the rectal tract. Human hands are actually quite small relative to the size of an elephant's rectal passage. To prove this, one only has to look at the size of elephant poop. But it was not enough to insert a hand and, ultimately, an arm. The rectal passage has also got to be cleared of any excrement, and this must be done manually as well. Eventually, the elephants learned to stand still for up to an hour, if necessary, while keepers cleaned out their rectal tract and inserted a thin tube into their vagina. Their only requirement was a steady supply of treats, their preference being the special biscuits we feed to monkeys, along with fruit and alfalfa hay (a special type of hay that they use only during procedures because it would otherwise be too high in protein). For some reason, our elephants loved monkey biscuits and would gladly hold still for as long as we were willing to pop the little crunchies in their mouths. The really good news was that, eating machines that they are, the elephants never got full.

The remaining problem involved getting the semen. Dr. Dennis Schmitt, an associate professor at Missouri State University and veterinarian at the Dickerson Park Zoo in Springfield, Missouri, has worked closely with elephants throughout his career, and he was the ideal candidate for this particular job. Dennis is a stocky guy, but not exactly a big man. Anyway, he doesn't look large enough to wrestle an elephant penis. He doesn't say much, but he radiates the calm confidence that characterizes so many people who work with elephants. Dennis had been an attending physician at thirteen elephant births by the time he started working with the Indianapolis Zoo. He had pioneered a different technique used to artificially inseminate elephants. This technique involves cutting a small opening under the anus and inserting the semen directly into the cervix. His technique works, and Dennis was the first to successfully inseminate an elephant (in this case an Asian elephant) in 1997, but his technique is invasive. It requires a surgical incision, which allows for the possibility of infection and requires some form of restraint, coupled with the use of a local anesthesia. Dennis was very excited to collaborate on our new noninvasive technique, which was voluntary on the part of our cows. They are not restrained and no anesthesia is required.

Dennis collects semen by having an assistant gently massage the bull's prostate, which is reached via the rectum, while he manually stimulates the bull's penis. In other words, he simply masturbates the bull. When the bull ejaculates, he collects the semen in what amounts to a giant condom. A bull's penis is several feet long, up to one and a half feet in circumference, and has musculature that allows the organ to be moved in any direction. It looks and moves like a giant snake. Dennis reaches into the holding area between the bars, places the condom over the bull's penis, and moves it back and forth until he ejaculates. The process takes more than a little courage. If the bull gets excited—and let's face it, that's the whole object of the exercise—he could easily crush one or both of Dennis's arms or even his entire body against the bars. (As an interesting footnote, the bull that Dennis perfected his technique on was named Onyx. Onyx later fathered the most popular animal in the Saint Louis Zoo, named Raja, but Raja was conceived the old-fashioned way!)

In addition to our regular staff, we also added Dr. Robert Hermes, a post-doc from IZW. He came to Indianapolis for several months to work with our staff and elephants. By Memorial Day weekend of 1998, we were ready. Our elephants were trained, we had practiced the insemination techniques, the Germans had flown over (thanks to the three-week notice given by an elephant's twin LH surges), a bull at the Kansas City Zoo had "volunteered" his semen, and our staff had a well-conditioned and cooperative female. Kubwa was going to be a mother.

Kubwa was what scientists call a "nulliparous" cow. In other words, she had never given birth before. In fact, she had never even had sex before. Kubwa was, in layman's terms, a virgin.

Early on the morning of Memorial Day, Dennis Schmitt began to masturbate Dale, a twenty-year-old bull who had been living in the Kansas City Zoo since 1994. After getting the semen from Dale, Dennis placed some under a microscope to assess the number of sperm cells in the sample and the extent to which they moved about in the seminal fluid. He found plenty of very active sperm cells. Dennis mixed the sperm with an extender solution made of fresh egg yolks, citrate, salts, and buffers, placed it in a chilled container, and rushed to the airport to catch a flight for Indianapolis.

As Dennis was boarding the plane, the crew at Indianapolis was just beginning to prepare Kubwa for the AI attempt. The elephants know when something is up. They could see the ultrasound machine being wheeled into the holding area and could tell by the excitement in the keepers' voices and the heightened level of activity and change in routine that something important was happening. Once Dennis was in the air and nearing Indianapolis, the keepers began to clean out Kubwa's rectal tract using their hands and warm water from a hose. The process was nearly complete as Dennis approached the zoo by car.

Thomas put on a special headpiece with an ultrasound screen mounted inside it. He would have to put his arm so far into the elephant to guide the insemination tube that he wouldn't be able to crane his neck around to see the screen unless the image was placed directly in front of him. The headpiece made him look a little like an alien from a cheap grade-B horror film, but it worked well. Thomas inserted the ultrasound probe while our staff, standing under the elephant, guided the endoscope and the attached semen tube into Kubwa's vagina. In about twenty minutes, the tube had reached the hymen, and Thomas had found the correct hole that led to the cervix and on to the uterus.

At this point, Dale's semen was inserted into the tube, pushed six feet along, and then deposited into Kubwa. The images from both the endoscope and the ultrasound were recorded, and all of our staff watched them on the screen as the procedure went along. Once the semen was released, it took only a few minutes to extract the scope and the ultrasound probe, and Kubwa was sent on her merry way, having eaten over fifty monkey biscuits. The Germans packed up their equipment and headed for home.

We continued to monitor Kubwa's hormone levels after our AI attempt. We were pretty sure that something was going on, but we didn't know we had a pregnant cow until July 24 when, during one of our regularly scheduled ultrasounds, we were able to confirm the presence of a developing fetus. In September, Thomas came back from Germany to do another ultrasound, and he got a wonderful picture of Kubwa's eighteen-week-old baby-to-be. An elephant baby as seen on an ultrasound doesn't look that much different from a human baby, aside from the trunk. The eyes are huge, the braincase is large in proportion to the body, and the four limbs are clearly distinguishable. In other words, it looks like a baby—and our staff reacted to the positive proof of Kubwa's pregnancy with unrestrained elation.

While they would have another seventeen months of waiting, they would not remain idle. On Halloween Day of 1998, the whole process was repeated, this time with Ivory, our youngest elephant. It would be another three months before Thomas could return to America to ultrasound Ivory, but when he did, he found a healthy three-month-old fetus. Two successes in a row meant that the technique was both effective and replicable. More importantly, it meant that we had a safe and simple way of helping elephants to reproduce without moving the animals from zoo to zoo. Their long-term future in North America had taken a dramatic turn for the better.

On an early Monday morning, March 6, 2000, that assurance became a reality. Kubwa's baby, a 201-pound girl, was born after a two-hour labor. Judy Gagen, the zoo's PR director and a passionate lover of elephants, was on the scene shortly after the baby was born, and it is her description that colors my memory. The staff had fretted, waited, and watched for the better part of two years. They had worked for ten years to achieve a live birth, and the barn, she said, had the feeling of a chapel. The quiet was unearthly. When people did speak, their voices were hushed and tinged with reverence. Perhaps birth is always so mysterious. Or perhaps two years of pent-up emotions couldn't be released with a silly high-five and a champagne toast.

Early on a Friday morning, August 4, 2000, a second calf was born, to Ivory. This time it was a strapping bull that weighed in at 252 pounds. They were the first two African elephants born via AI, and Kubwa and Ivory were the first two virgin elephants ever to give birth. It may not be politically correct to term them immaculate conceptions but, to those involved, they did seem like two "little" miracles.

11

The Stradivarius of Birds

There is some debate about how many violins made by Antonio Stradivari survive to this day, but seven hundred instruments is a pretty good guess. They vary greatly in value, but they can fetch up to $3.5 million at auction. There is no debate about how many Guam kingfishers survive to this day. As I write this, there are fifty-eight birds left alive. I have no idea what you'd have to pay to own one, but my guess is they are worthless. There is simply no economic demand for them.

The Guam kingfisher is a beautiful bird with an incredible story. Properly called the Micronesian kingfisher, it was found only on the island of Guam. At some point during or immediately after World War II, a cargo ship accidentally brought some vicious stowaways to the kingfisher's island paradise. The stowaways were brown tree snakes, a mildly venomous native of New Guinea. The tree snakes had no natural predators on the island and their population size went from a couple to perhaps ten thousand per square mile in just a few short years. And they ate almost every bird on the island, including our beautiful kingfishers.

Let me give you an idea of how many snakes "ten thousand per square mile" really means: the common cause of power failures in Guam is the weight of brown tree snakes hanging on utility lines.

By the time we got to the Guam kingfisher, there were only twenty-nine birds left. By 1990, we had built the population up to sixty and it has pretty much stayed right around that number ever since. Today, eight of those birds live at the Saint Louis Zoo—three breeding pairs and two chicks (or about 12 percent of the world's population). We don't put them on display. They're simply too valuable to us, even if they don't have any value on the open market.

A Stradivarius can go for upwards of $225,000 an ounce, and people really do pay that much! They are revered and much sought after. The Smithsonian keeps them in its collections, and we as taxpayers underwrite their care. They are considered art. As one enthusiast wrote, they dare to "challenge the ingenuity of God's own designs."

If scarcity equates to value, then the Guam kingfisher should be even more valuable than a Stradivarius. There are fewer of them (and they weigh less, too). I'll grant you that they are not considered art, nor do they challenge God's own designs. On the other hand, they are exquisitely beautiful and they don't merely challenge God's own designs. They *are* God's design.

I think I understand why people care so much about a Stradivarius violin and its sweet song. What I don't understand is why people don't care just as much, if not more, for a Guam kingfisher and its plaintive call. In the end, I can't figure out why people who care for one don't automatically care for the other, but I think the problem might be that people just don't know enough about the Guam kingfisher. Like so many things in our world, it's not about the law of supply and demand so much as it is about creating demand through slick marketing.

You can't sell what people won't buy, and zoos are, to a certain extent, like any other business. The good work that we do costs money. If we want people to visit, this place has to provide a fun, marketable experience. On the other hand, there is nothing fun about extinction. Let's face it, it's a lot easier to get people to fork over two bucks for a balloon than it is to get them to fork over two bucks to help save lemurs in Madagascar.

Zoo directors have been impaled on the horn of this particular dilemma since the very beginning. We're half serious conservationists (largely by personal preference) and half marketing weasels (largely because of our actual jobs). It is, at least for me, one of the most difficult parts of what I do. The whole thing was brought front and center for me shortly after I first started working in St. Louis.

When I arrived, the first thing I did was take a careful look at the zoo's finances. We had just finished a $70 million capital campaign and had added some wonderful new exhibits. The problem was (and still is) that they are all very expensive to operate. For example, in 2003 we opened a huge new penguin and puffin exhibit at a cost of about $8.2 million. The exhibit features about one hundred birds in a giant, domed building. The building is cooled to about forty-five degrees, and the birds can swim up to within inches of the public. The visitors simply love it. The icy air, combined with the wild antics of the birds, the smells, the cold water that the birds splash on our guests, and their raucous calls all make the experience very memorable. We could have put the birds behind a wall of glass, and they still could have swum close to our guests, but it wouldn't be the same. With no barriers between the birds and the people, you get a truly immersive exhibit. Feeling the cold, smelling the fish, and hearing the birds make it as close to a visit to the Antarctic as you can get.

The thing is, it's expensive to chill that huge volume of air. Plus, with people being able to reach over the glass and touch a swimming penguin, we have to add security to monitor the exhibit whenever we're open. This is partly for the sake of the birds, but partly for the sake of the visitors. The birds have sharp beaks, and it's relatively easy to confuse a little finger with a tasty fish—and they go through about one hundred pounds of fish per day. That's a pretty steep restaurant tab. Finally, you need more professional staff to care for that many new residents. To make a long story short, this exhibit, and almost everything else we added to the zoo, made this place more expensive to operate.

On the other hand, we didn't increase our attendance enough to really increase our operating revenues. The problem wasn't that the new exhibits weren't exciting enough to draw more visitors. The problem was (and still is) that there simply isn't any place for them to park their cars. The Saint Louis Zoo draws about 3 million visitors a year. Most of them come in May, June, July, and August. During those months we fill our parking lots to capacity about fifty times. We have two lots, and there are about twenty-four weekend days during the summer. This means that our lots close on just about every nice weekend day. It really doesn't matter that people are nuts about seeing our penguins. If they can't find a place to park when they get here, they can't come in. If they can't come in, we lose parking revenues, food and beverage sales, gift shop sales, and so on. We added exhibits, not parking. That means, in effect, we added costs and not revenues. It wouldn't take a rocket scientist to realize that this would, sooner or later, catch up with us. As I looked at the books, I figured that by the time we finished all the new construction we had planned for the future, which would take about three more years, we would be in the red from an operating point of view.

There are really only a handful of ways of dealing with a problem like this. You can decrease expenses, which is very hard since most of the new expenses are related to all the wonderful new exhibits, or you can increase revenues. Since we don't charge admission, this means we have to charge more for food, gifts, and all the other things we do to separate visitors from their money. But there's only so much you can charge for a hot dog before people will say, "How about we just stop at McDonald's on the way home?" Face painting is certainly a luxury that many families can forgo, and those balloons will probably break or float away in a moment of inattentiveness, so they're not really a life necessity, either.

Assuming that prices are already set pretty much at what the market can bear, the simple solution is just to add more revenue producers. My idea was to add an endangered species carousel. The one I had in mind was a sixty-four-animal, hand-carved and hand-painted classic, custom made in Mansfield, Ohio. They're expensive to buy, but they're beautiful. Many zoos have added them in recent years, and I thought, "Hey, this could keep the wolf from the door for an extra year." My idea, though, wasn't greeted very warmly. In fact, you would have thought that I had suggested robbing visitors at gunpoint.

The first criticism was that I would ruin the ambiance of the zoo. "This is a bio-park," they said, "a serious conservation organization. Carousels have no place here." Others felt that it would cheapen or demean the real thing. "You can't have people riding a hand-carved Siberian tiger with a jeweled saddle," they exclaimed. "That communicates the worst message—that people are supreme over animals and that they're here for our entertainment." Several board members, proud of our history of being a free zoo and, as such, available to virtually anyone regardless of whether they're rich or poor, felt that people who couldn't afford to ride would be upset.

Of course, the people in the marketing department (who are responsible for keeping the attendance high), along with the people in the finance department (who have to balance the budget and pay the bills) thought the idea was absolute genius. The people in the development department were just sort of "okay" with the concept. They like raising money and felt that the carousel could be funded from charitable gifts, but they had already raised quite a bit of money and thought maybe they should give our donors a little break, rather than jump in with a new project when we were still three years away from finishing all of the other projects in our last capital campaign. Suffice to say that there was a striking lack of consensus.

Luckily, there's an unwritten rule that says you can do almost anything you want during the first year you have a job. It's an official honeymoon. I took advantage of it and, within a year (thanks largely to a wonderful zoo donor named Mary Ann Lee) we had a new carousel. We wound up raising almost three times as much as it actually cost to build, we offer it free to the public every morning for the first hour we're open (so you can't argue that it disenfranchises poorer visitors), and it will make a couple hundred thousand dollars a year, helping to underwrite what we do. All of the naysayers are quiet now, and everyone is happy except me. It's not that I hate the darn thing, the fact is, I love it; it's just that I hated having to do it.

It's not just the carousel—it happens with almost everything around here. Take, for example, something as simple as exhibit graphics. One of our projects, completed in the spring of 2005, was the creation of some marvelous outdoor exhibit yards for our chimps and orangutans. When I got here, we were just beginning the process of figuring out exactly what we wanted to say to our visitors. In effect, we were wondering what stories we wanted to tell. There are plenty of things to say about these wonderful creatures, and it shouldn't be hard to get people to identify with these, our closest relatives, in a profound and real way. For me, it provided a chance to get the staff to buy into my ideas about the power of stories and story telling. The problem was that there were just too many stories to pick from, and what was worse, the most important story we could tell would probably be almost too horrifying to present.

First let me tell a nice story about our apes. Maybe that will help put the more horrifying one in context. We have a chimp named Cinder who has a disease called alopecia areata. She was born here at the zoo in 1994, and about five months after her birth, she began to lose every single hair on her body. This disease often strikes humans, and dermatologists continue to look for a cure. It strikes men and women and does not discriminate on the basis of age. Some victims lose all their hair at once, but for some, it takes awhile. Heredity appears to be a factor in about 20 percent of the cases. Alopecia areata doesn't hurt or itch, it's not contagious, and it does not lead to any other physiological problems. But psychologically it can be devastating.

In July 2002, St. Louis was the host city for the National Alopecia Areata Children's Camp. One hundred fifty children afflicted with alopecia visited the zoo as part of that camp, and every one of them spent at least half an hour looking at a chimp that had the same disease that they had. What fascinated them, and what fascinates me to this day, is that Cinder, while she looks totally different from all the other chimps in her troop, is not treated any differently than any other chimp. People sometimes seem to care more about appearances than substance. Chimps don't. Thankfully, it seemed that the kids visiting Cinder that day already knew that lesson. I overheard a little boy named Corie talking to a reporter from the newspaper, who was covering their field trip to the zoo. He said he had a whole collection of bandannas to match his clothes, but that he wore them as much to put others at ease as he did to cover his little bald head. "It doesn't matter at all to me or my friends," he said. "I still am who I am inside."

I don't think that Cinder and our other chimps taught any of those kids anything they didn't already know about being human, but they sure could teach the rest of us something.

I think that's a nice story, but I don't think it's the story that we should be telling at the zoo. I think the story that we should be telling is of the unsustainable slaughter of chimps, pygmy chimps, and gorillas in the wild—what we have come to call the "bush meat crisis."

There are three subspecies of gorilla. Right now, the mountain gorilla is the rarest, with some 600 animals left alive. They are found only at relatively high altitudes, mostly between 5,400 and 12,000 feet above sea level, and they exist in the highlands of the Congo, Rwanda, and Uganda. The other two species, the eastern and western lowland gorillas, used to be relatively abundant. But in the Congo, over the last three years, the population of the eastern lowland gorilla has gone from about 17,000 individuals to less than 1,000—a drop of over 90 percent! In Cameroon, the Central African Republic, Gabon, the Congo, and Equatorial Guinea, the strongholds of the western lowland gorilla, the population is still pretty high—about 80,000 animals. But they are dying at a rate of over 100 per week, and at that rate, they will be extinct in ten years. Let me try to put this in perspective. If every gorilla in the world were to attend a home game at the University of Michigan, the stadium would be less than 80 percent full.

What's driving the gorilla to extinction is the same thing that is driving chimps and pygmy chimps to extinction: humans are killing these, their closest relatives, and eating them.

The people of Africa have always killed wild game for food. The common term for wild game is *bush meat*. Gorillas, chimps, and pygmy chimps have been on the menu since humans began hunting. The problem is that, at the present time, people are killing them at a rate far higher than at any other period in history. There are several reasons that the rate has skyrocketed, but the most important reason, at least in the Congo, has to do with cell phones.

Yep, cell phones. That ringing at a nearby table while you're trying to enjoy a quiet meal at a restaurant may be a minor irritant to you, but to the eastern lowland gorilla it is a death knell.

The electronics industry is the biggest user of a mineral called coltan. It's used in computer chips, and the average cell phone requires quite a bit of it. Coltan is found in several places around the planet, including Australia and the extinct volcanoes of Greenland, but it is expensive to mine coltan in developed countries. It is cheap, however, to mine coltan in a place like the Kahuzi Biega National Park in northeastern Congo. This national park, which covers some twenty-five hundred square miles, was the birthplace of gorilla tourism in Africa. The park had an abrupt reversal of fortune, however, beginning in 1994 with the genocide in Rwanda. Refugees flooded into the Congo to escape the carnage and, shortly after that, armed militia groups also moved into the park. At one point there were nine different armies occupying the park, and all of them were funding their operations by mining coltan. In fact, as I write this, 90 percent of the park is in the hands of warlords. Says park warden John Kahekwe, "these soldiers live alongside frightened villagers who have nothing to eat. They are told that they must scrape the earth looking for coltan." Others, perhaps as many as fifteen thousand people, have moved into the area to mine coltan, in addition to the villagers who have long lived in the park. These outsiders come without food, cattle, or access to crops. They mine coltan for the warlords, who smuggle the mineral out. The miners may make a few cents a day, but on the international market, coltan sells for anywhere between $75 and $150 a pound.

The miners eat birds, insects, flowers, and grass, but their main source of protein is meat from chimps and gorillas. The park is overrun with men carrying AK-47s, and the apes have nowhere to hide. Their bodies wind up in cooking pots and their heads, hands, and bones wind up as trophies. Photos from the public marketplaces show huge piles of heads and hands available for sale.

Even where coltan is not mined, like the strongholds of the western lowland gorilla, hunting for bush meat by professional hunters is dramatically reducing the population. The situation is not that much different from the crisis that confronted the American bison of the Great Plains. Habitat loss resulting from increasing the acreage devoted to farming, plus a dramatic increase in accessibility (on the Great Plains this was a result of the expansion of railroads, but in Africa it is more because of logging roads), plus a market for meat and hides almost spelled the end of the buffalo. Now the same thing is happening to apes in Africa.

When I tell people about this, they're amazed. They say they had no idea this was going on in Africa and want to know how to stop it. But telling the story well means displaying some pretty gut-wrenching photos of the carnage. People don't come to the zoo to be depressed by the sad plight of gorillas and chimps in Africa, goes the logic. The story is horrifying and shocking, not suitable for young children. Who wants to come to the zoo to be depressed and disheartened? It's just not good marketing. But we must do it anyway. We must make it as gut-wrenching and graphic as we can.

We have something of the same dilemma with our shows and demonstrations, especially here at the Saint Louis Zoo, where shows were historically a big part of the zoo experience. Beginning in 1925, the director of the zoo, the legendary George Vierheller, created a series of live animal shows that, in Vierheller's words, made a circus out of the zoo, "rather than a collection of musty specimens." The first "really big show" was the Chimpanzee Show, which opened in 1925 on a stage that was built as part of a new primate house.

In her book *Wild Things: Untold Tales from the First Century of the Saint Louis Zoo,* Patricia Corrigan devotes a whole chapter to the shows. A photo from the archives indicates how elaborate the sets were, with five or six chimps, all in costume, performing an elaborately staged show. At one time, for example, we had two chimps that were trained to box. Battling Billy Busch and Sock'em Sammy Green went four rounds in each of seven daily shows in an act that included tumbling, swinging, and jumping. As Vierheller described it, "Both of the chimps entered into the fight with much gusto and showmanship. They bared their teeth at each other, and often squealed when a foul was made or a sore spot hit. Sammy was supposed to take the count, but once, he became tired of being the loser and deliberately knocked Billy out of the ring."

The chimps were trained to ride ponies and, in one case, a Great Dane. They rode tricycles, bicycles, motorcycles, and unicycles, and some of them developed their own followings among the general public. Pancho, who became known as "Mr. Mischief," was a case in point. According to Vierheller, "He often unexpectedly trips the stilt walkers, unseats the bicycle riders, drives his jeep the wrong way to cause a crash, and tickles the ponies to scare the jockeys. After his misdemeanors are done, he always runs to the center front of the stage, grins in chimpanzee style, and waves to the crowd." By the early 1950s the equipment in the Chimpanzee Show expanded to include "wooden floats, a scale-sized old Maxwell car, a crazy car that leaps and bucks, Jeeps cut to size, motorcycles, fancy bikes of all kinds and shapes, unicycles, swinging giant trapeze combinations, big balls, teeter boards, roller skates, and stilts." There was also a merry-go-round and an orchestra platform with a variety of instruments for the chimps to play. The show began to decline in 1977, with the death of longtime trainer Mike Kostial, but was not abandoned for good until 1982.

It's hard to imagine a modern zoo presenting such a ridiculously anthropomorphized show today, but 1982 wasn't all that long ago. And it wasn't just the chimp show that used animals for entertainment in this fashion. Pat Corrigan, drawing heavily from an archival manuscript prepared by Jim Alexander here at the zoo, says, "The first Elephant Show took place in the summer of 1937, in the yard of what was then the Camel Barn. The elephants

Vintage scene of Saint Louis Zoo's old chimp show. SAINT LOUIS ZOO

danced a bit and played the drums. In 1938, a new act was added. The elephants drank from metal bottles and then acted tipsy. The Anti-Saloon League protested that this behavior was offensive, but the Zoo Board voted five to four to keep the routine." The Elephant Show always seemed to feature something egregious. The 1945 show had a barbershop routine that featured one elephant "shaving" the other one. During their tonsorial interlude, a fight would break out during an elephant craps game going on in the back of the barbershop. The show ended with one of the elephants chasing the other off the stage with the straight razor. It wasn't until 1991 that the zoo's elephant show began to shed the most overt of its anthropomorphic stunts—and that was the last year the Elephant Show was presented.

We also had a Lion Show that included up to twenty-five lions in the ring at the same time. Vierheller loved it, and Corrigan quotes him as saying that it was "a breath-taking display of bears, tigers and about seventeen trim female and regal male lions going through formations and stunts with precision-like ease. The bears dance to the strains of 'Three O'clock in the Morning' while they drink milk from a bottle, ride a hobby horse, and roll off the stage in scooters; the tigers take to their perches as they are bid, or whirl with dizzying speed at a given cue; the lions pyramid, walk and dance on their hind feet, jump through paper-filled hoops, hold a rope swing for their trainer, roll balls, and fight off their trainer's stick with convincing ferocity."

We still do a sea lion show to this day, but it doesn't try to make sea lions into people. It showcases natural behaviors and, like many shows that are still done in the zoo world, has a strong conservation message. But it's still a show, and I'm not sure it really has a place in the zoo of the future. Year after year, however, it comes out on the top of every survey we do. People love it and gladly part with two dollars to see it. The show enhances the zoo experience and generates over $150,000 a year in revenues.

Still, the same criticisms that you can level at the carousel can be leveled at the sea lion show. One can say that it doesn't belong in a modern zoo, that it cheapens our reputation as a serious research and conservation institution, and that it demeans the animals by forcing them to perform for our amusement (even if it is educational).

Deciding which animals to have in our collection raises the same sort of questions. On the one hand, you could argue that zoos have absolutely no business keeping an animal like, say, a polar bear. They're not the least bit endangered, with an estimated population of twenty-six thousand. There are no pressing research questions that could only be addressed by using a captive population. Sure, they demonstrate some fascinating adaptations, but you could probably display an Arctic hare and get the same points across. On the other hand, people love them. They come to see them. If they connect with polar bears, you might actually get them to care about Malayan sun bears. There are only some six hundred to a thousand of them left. In fact, there are probably more giant pandas left alive than there are sun bears. Sun bears are fascinating animals to watch. They're incredibly active and agile, they're curious, and, though I hesitate to use the word, they seem almost *personable*.

Clearly zoos ought to phase out polar bears and bring in sun bears. The problem is that people don't care about seeing sun bears and they do care about seeing polar bears. Also, "phasing out" can really only mean one thing—waiting for the bears to die a natural death. We can't return them to the wild. But by the time the population of polar bears in human care drops to zero, the sun bear will probably be extinct.

It's the same thing with zebras. There are about six hundred thousand plains zebras living in Africa. They aren't the least bit endangered. But the plains zebra (more precisely, Burchell's zebra) is only one of three species of zebras. The other two are the mountain zebra and, perhaps the most striking of all zebras, the Grevy's zebra. Grevy's zebras differ from the common zebra in several ways. First, they are found in a relatively restricted range—southern Ethiopia and northern Kenya. They look different from the common zebra as well. They are noticeably taller and heavier, with a massive head and larger ears, and their stripes are much closer together. Unlike other zebras, they do not form permanent herds, and the males require an extraordinarily large territory to breed—in fact their territories are the largest known for an herbivore. They are uniquely adapted to fit between the arid zone occupied by the wild ass and the relatively moist environment of the plains. Although they require less water to survive than their cousins on the open plain, they are less tolerant of the cold.

In the fall of 2002, we sent one of our curators, Martha Fischer, to Ethiopia and northern Kenya to survey the population size of the Grevy's. What she and her colleagues found was shocking. The preliminary results of the survey showed that the population of Grevy's in Ethiopia had declined more than 90 percent over the last twenty years. Between 1980 and 2002, the population dropped from 1,500 zebras to 110. Even if you add in the Kenyan population, the total number left alive would probably not exceed a couple thousand. Martha, along with the other members of the survey team, which was headed by Stuart Williams, a Conservation Fellow here at the zoo, spent weeks engaged in aerial and ground surveys, camping in dusty wallows and eating sardines and spaghetti. They were constantly surrounded by heavily armed men, and Martha once told me that it seemed like every male over the age of fourteen had an AK-47.

As is the case with Africa's great apes, humans are the immediate threat. The herdsmen of Ethiopia view the Grevy's zebra as competing for the scarce water. Despite taboos against it, people admit to eating Grevy's whenever they can, and they have always attributed medicinal properties to zebra parts, the same way people do with rhino horns.

The team's survey was difficult and potentially dangerous, and the results were highly discouraging. In fact, the whole time Martha was gone, her colleagues here at the zoo were worried sick. When she finally returned safely, we decided that the zoo had to buy a portable satellite phone so that we wouldn't have to worry so much. "Forget about the whole coltan thing," went the logic, "life is a series of compromises."

Zoos simply aren't really working with the Grevy's zebra to the extent that we could. Globally, there are 647 Grevy's found in 132 zoos. Only 44 zoos in North America have Grevy's zebras, and the total population on our continent is only 230. Conversely, there are currently over 1,060 plains zebras, which are far less useful from a conservation point of view. We should have known better. Anything that is native primarily to Ethiopia and is good to eat is probably going to be in trouble sooner or later. Ideally, we would replace many, if not most, of our common zebras with Grevy's zebras.

In the case of the Grevy's, we wouldn't really have to create demand to see them. From the public's point of view, they'd still be seeing zebras, and with the Grevy's, it would be easy to convince people that they're seeing an even more special species. But with polar bears and sun bears, we'd have our marketing work cut out for us. However, I still believe that we could create the same kind of demand for sun bears as we already have for their arctic cousins.

Of course, if we could create the same kind of demand for the Guam kingfisher as we have for the Stradivarius, then we'd really have achieved something worthwhile. Our kingfishers would be worth a fortune. I mean, imagine this. I'll let you see the Guam kingfisher for a mere one thousand dollars. Okay, not much of a deal, *but* for one *million* dollars, I'll let you own one. Just like people buy Stradivarius violins and then let the Smithsonian keep them, you'll have to let us care for the bird. Still, it will be yours. Rarer than a

Stradivarius, sweeter than a symphony, more delicate than any concerto, this beautiful bird can belong to only a lucky few of the wealthiest of people. The zoo could be rich! Heck, we have more Guam kingfishers than any other zoo in the world. We could corner the market! With some luck, coupled with some marketing genius, we could soon be referring to Strads as "the kingfisher of violins."

Until we can manipulate public demand with the skill and aplomb of, say, a ketchup manufacturer, zoo professionals will remain on the horns of the marketing dilemma. We'll constantly be compromising so we can stay in the business of conservation and public education. I guess it could be a lot worse. Just think of the rhinoceros. Their horns *are* their dilemma.

12

We Haven't Lost Anybody (Yet)

One of the best things about being a zoo president involves traveling to some of the world's most exotic places. As a child, I dreamed of seeing the world and living the life of a nineteenth- or early-twentieth-century explorer. To a certain extent, I have. I have rafted the lower Zambezi River, floated in a hot air balloon over the Serengeti, climbed to the top of Machu Picchu, swum with black-tip reef sharks in the Bahamas, hiked the dunes of the Namib Desert, snorkeled with the great Pacific rays of the Galapagos Islands, kayaked with river dolphins in the Amazon, fished off the Skeleton Coast, photographed hummingbirds in the Costa Rican cloud forest, and captured rhinos in the Kruger. All of this is considered work, but it's hard to see it that way. In fact, I used to say that, even if it is work, I'd still pay somebody to let me do it. I stopped saying that, though, when my wife pointed out that I wasn't likely to get another raise if word got out that I was having so much fun.

Although I occasionally travel for scientific purposes, most of the time it is because the zoo sponsors international trips as a way of getting donors (or potential donors) educated on issues in conservation. There isn't a better way to get people to fall in love with wild places than actually taking them to those places. The catch is that they are indeed wild places, and things can, and do, go wrong. As a result, our unofficial travel slogan is, "We haven't lost anybody (yet)." I guess you could say we're appealing to a niche market. We have certainly come close to losing a few folks, but for the most part, people think that they're in a lot more peril than they really are.

For example, on our first trip to Africa, my wife and I found ourselves spending our first night out in the bush in a small tent camp in the Okavango Delta of Botswana. Botswana represents Africa as it was hundreds of years ago. The country is quite large (about the size of any three midwestern states put together, or roughly the size of Texas), but nobody lives there. The population was reported at 1,480,000 in 1998, or about 6.4 persons per square mile. Compare this to India, with 735 persons per square mile in 1998, or Kenya, a country that has some amazing wild places, yet still had 126 persons per square mile in 1998.

The Kalahari Desert covers much of the southern and western parts of Botswana, and the Makgadikgadi Saltpans and the Okavango Swamps, the world's largest inland delta, dominate the northern part of the country. Here the mighty Okavango River flows onto the hot dry plains of Botswana, finishing its 2,700-kilometer journey across Africa. The river simply dies in the sand. It is not good land for cattle because of the presence of tsetse flies, and it is difficult to farm because of the annual inundation of the river, so the land and the wild animals that live upon the land remain much the same as they always have.

Our first trip there was with a group of sixteen zoo directors. We knew we would be leading trips on our own, and we wanted to get some experience before taking a trip where we were actually in charge. After flying twenty-two hours straight, we arrived in Johannesburg in the morning and spent our first day and night in a lovely hotel in the city, recovering from an awful plane ride. Early the next morning we boarded a smaller plane bound for Maun, the only town close to the delta with an airport, and from there, transferred to six-seat Cessnas, flown mostly by young bush pilots.

The planes fly low, and herds of zebra, elephant, and wildebeest can be seen easily from the window. The pilots navigate by landmarks only they can see, and the airstrips are just grassy fields. On arrival, the pilot buzzes the field to scare off the game, but landings are still something of an adventure. Six hours after leaving the hotel in the city, you can find yourself in the middle of nowhere.

The camps themselves tend to be luxurious. Sure, you're in a tent, but it's a large tent, often raised a few feet off the ground to discourage unwanted visitors, and covered with a thatch roof to provide shade from the noonday sun. The interiors feature hot and cold running water, claw-foot tubs and flush toilets, electricity during the mornings and evenings, sitting areas, and beautiful wooden verandas facing out to the open plains. Almost all of them have, in addition to the regular indoor plumbing, wonderful outdoor showers, although standing naked before a herd of curious impala takes some getting used to. There is almost always an open-air wooden lounge with a dining area and, of course, a full bar, and there is always a fire pit where stories of the day are retold into the night.

Arriving around two o'clock in the afternoon on our first day, we had plenty of time to unpack and ready ourselves for our first game drive, a trip across the plains in a modified Land Rover. Each vehicle seats up to nine passengers on three bench seats, each seat about a foot higher than the one in front of it in order to maximize the view in all directions. The seats are higher than the sides of the vehicle, and there's no top, but the drivers proceed slowly enough that there's no sense that you might be pitched out over a large bump. On the other hand, you feel more than a little exposed when you first pull up next to a pride of five or six lions. Sitting ten feet away from a full-grown lion, your first realization may be how magnificent these animals are, but your second realization is probably more along the lines of, "My God, they could kill me in one bound!"

We saw so much that was new to me on the first day that my head was swimming. The other zoo directors thought Melody and I were a riot. They knew so much more about African wildlife than either of us and didn't mind teasing us relentlessly about it either. Their favorite pastime was inventing barely plausible names for questionable birds. It went something like this. If they were pointing in my general direction, they'd say, "Oh, there's a wide-eyed, white-chested administrator bird!" All the zoo directors were men, and I was the only one who had brought his spouse, so we got the "honeymoon tent." I don't think it was any nicer than the others; it was just located farthest away from the main camp!

As night fell, we settled into bed and listened to the lions as they hunted in the darkness. Lions hardly ever roar unless they are in great distress. When they call to each other in the night, the sound is more of a low, sustained grunt that carries a surprising distance. We lay awake listening, and the calls came closer and closer to the tent. Suddenly I felt my pillow being ripped out from under my head. In the next instant my wife had also retrieved the extra pillows from the chest at the foot of the bed and began packing them tightly around her body. "What," I asked, "could you possibly be doing?"

"Okay," she replied, "when the lion rips through the tent and tries to eat me, it's going to get nothing but feathers at first, so it'll probably just start on you."

"That's ridiculous," I replied, "to say nothing of being mean-spirited."

"Say what you will," she concluded, "I have a plan. You, on the other hand, don't even have a pillow." We laughed ourselves to sleep, but awoke to find lion prints in the soft dirt alongside the tent.

We always seem to wind up in the honeymoon tent. We took a group of zoo supporters on a tent safari in Kenya and pitched camp alongside the Mara River. After dinner, a guide escorted us back toward our tent, carrying, as always, a rifle. As we neared our tent, perhaps some twenty yards away, an elephant trumpeted in the darkness. We never saw the animal, but it sounded loud enough to have been within feet of us, standing in the dense under-brush. My wife looked like one of those cartoon characters whose legs spin around in a little circle before they suddenly take off. But she moved so quickly that the guide could only stand in amazement until something clicked in his brain and he started to run, too. Unfortunately, he had the rifle. I was the last one to make the safety of the tent. As they say in Botswana, "I don't have to outrun the angry elephant—I just have to outrun you."

If you listen carefully to the guides and do as you're told, you'll probably never be in any true danger. Certainly, predators like lions and leopards move through the camp at night, but other than some extraordinarily well-publicized exceptions like the lions of Tsavo, they would never try to rip into a tent. We do tell people to never, ever leave the tent at night, and that means you'll stay safe. On the other hand, we have had some brushes with disaster, usually when the guide pushes things just a little too far.

Our favorite guide in all of Africa is a man named Patrick Boddam-Whetham. He is impishly handsome, cultured, and a brilliant conversation-alist, making him very popular with the women on a trip. He is also a man's man. You get the feeling that he could survive in the bush with nothing but a sharp rock and a stick. Patrick was born and raised on a farm in South Africa and went to Rhodes University, where he studied geography. He has spent almost all of his working life guiding safaris throughout southern Africa, spending most of his career with an outfit called Wilderness Safaris. He rose up through the ranks of the company and is now in charge of the company's operations in Zululand, but because we've known him for many years, he will take time off to guide for us once or twice a year.

Wilderness Safaris does not "own" camps in places like Botswana in the traditional sense. Instead, they pay a commission to the government to operate a camp for a fifteen-year period. They send in crews to build the open-air lodges, erect the tents, build the kitchens and other support facilities, and make trails through the surrounding countryside. At the end of the fifteen years, they must rebid on the territory. As a consequence, it is very expensive to operate, but the revenues from ecotourism make the preservation of wilderness areas economically viable to the government. They also have a very aggressive program of training locals to work in and manage the camps, providing careers to many people who would otherwise have very few career options. Having Patrick along when we stay in a camp pretty much ensures that things will go well, but even Patrick occasionally makes an error in judgment.

The worst one I can remember occurred when our group had a free day at Victoria Falls. Called "Mosi-oa-Tunya," or "the smoke that thunders," the falls are caused when the mile-wide Zambezi River plummets 420 feet down into a narrow chasm created by a fracture zone in the earth's crust. The Zambezi separates Zambia from Zimbabwe, and the river actually splits into five separate falls as it flows down into the chasm. Farther downriver from the chasm, the sheer canyon walls tower up to 700 feet above the water.

The force of the water creates some of the most amazing rapids in the world because the chasm is so narrow and the river above the falls so wide. They are considered Class V rapids (Class VI "involves serious risk to life" but can be run), and the rapids extend some twelve miles from their beginning near the base of the falls.

Victoria Falls has some of the best shopping in southern Africa, but most of our group had already visited all of the shops and the open-air market by the time our last day at the falls dawned. We had until six o'clock in the evening open, so we offered the members of our tour several options, including a daylong horseback safari, an elephant-back safari, and a whitewater rafting trip. Three of our group picked the white-water rafting option. I had asked Patrick if it was safe, and he replied, in his Australian-sounding Afrikaner accent, "Oh sure, piece of cake." Still, I figured I ought to accompany our little group, just in case.

The three people who decided to go rafting were Beth Cate, a Harvard-educated attorney who was at the time about thirty-five years old and in great shape, and John and Cynde Barnes. John was probably in his middle sixties at the time, while his wife was in her early forties; she is the kind of woman whose zest for life is apparent in everything she does. They had some experience riding white water on the Colorado River and thought they knew what they were getting into.

Our first clue that this might be more of a young person's game occurred when we met in the morning for our safety lecture at a nearby hotel. We had been told not to bring any money or valuables and to leave our good cameras behind in favor of waterproof disposables. We were dressed appropriately, but when we looked around the room, it was clear that the other forty-odd rafters were all in their midtwenties. We looked the part, but I wasn't sure that we had, in the parlance of the American space program, "the right stuff."

My concerns increased a little when, after handing out the safety equipment (a helmet and a life preserver), the river guides began to discuss what to do if (and when) the raft flipped in the rapids. "Make yourself into a little ball if the vortex is trapping you below the water," they said, "and count to ten. If the river hasn't spit you out of the vortex by then, you should count to ten again. Repeat this process as long as necessary."

After the lecture, we loaded into some open trucks and made our way to the gorge. We got out of the trucks and began to walk down the long trail to the river's edge. This is when I should have had second thoughts. It simply never occurred to me that, if we had to walk down the gorge, we would also likely have to walk back up when we were finished. But the trail down wasn't that bad—it was early in the morning after a good night's sleep, and we were only taking the half-day trip, so we would have plenty of time to walk back up and catch our train. I was still thinking, "piece of cake."

At the edge of the river, we got into our rafts and practiced paddling along with some of the basic techniques required to keep the raft going through the rapids. The guide, who sits on the back of the raft and steers the giant inner-tube-style raft through the rapids, does most of the work. The skill comes in picking the line for the boat. All the passengers do is throw their weight forward at critical times to keep the boat from flipping upward and over. Everything else is up to the guide and the river gods.

After ten or fifteen minutes of practice, the current brought us toward our first major cataract. They all have appalling names like Devil's Toilet; I've forgotten the name of the first rapid, but I do remember cheerfully waving to one of the rafting company's employees, who was videotaping our first challenge. Seconds later the world exploded. The raft tried to fly three ways at once, our group forgot everything they had been told, and our focus seemed to dwindle down to one basic imperative—hang on for your life. The boat flipped high, one or two paddles went skyward, and Cynde disappeared for a very long time while her husband frantically called her name.

She finally emerged from a river boil, having counted to ten three times.

She was well behind our boat, which was floating upside down in the placid reach below the rapids. At the time, I was concerned for her and upset that her husband had been so frantic. I also have to admit that I, for one, was having a very good time. Beth was too, but then she hadn't been trapped under water for half a minute or spent any anxious moments searching the rapids above for signs of her spouse. In fact, her spouse was in Bloomington, Indiana. Cynde was picked up by another raft; we righted our boat and caught our collective breath. John allowed that their rafting experience in Virginia wasn't quite at the same level, and I asked the guide just exactly what Grade 5 meant. "I think," he said, "that it means there's serious risk of injury."

"Great. How often do you have injuries on these trips?" I asked.

"I guess somebody dies every couple years or so," he replied, "not too bad, considering."

Now that's the kind of information you really need to get *beforehand.* Our day turned from being a happy lark to a live version of one of those grim survival television shows. As we came up to the next rapids, we were no longer waving to the videographer stationed on the bank. Instead, we were giving him the proverbial finger—another case of the displaced aggression we find so often in the animal world.

I don't recall which of the nine rapids flipped our boat the second time, but I do remember holding on to the upside-down raft and reaching for Cynde so she wouldn't get stuck underwater again. Our fingertips touched in the churning, foamy water, and she used her sharp fingernails to slowly inch her way up my arm and toward the safety of the boat. By the end of the morning, we were exhausted.

Things began to look up when we pulled over to the riverbank for lunch, though. The terror of the rapids was past, we were all in one piece, the food was good, and the hot sun soon dried our clothes and raised our spirits. About half of the people on the expedition were only taking the half-day trip, and slowly we began to meander over to the trail that would provide our means of ascent from the gorge.

The trail wasn't bad at first, but it gradually grew steeper and rougher. In places, they had made primitive ladders from branches with the rungs attached by bailing wire. We reached for roots and slipped on the damp soil. We found ourselves, in places, literally clawing our way out. Away from the cool river, the heat was growing intolerable, and John began to tire.

What none of us knew, including John, was that he was mildly diabetic. This probably wasn't the best place in the world to figure this out. We had a seven-hundred-foot vertical ascent before us, there was no one around to help, and John was suddenly weary beyond imagining. "Just leave me," he said. "I'll make my way up later."

"Not a chance," I replied. "We're in a national park and there are things that can eat you here. We'll wait for as long it takes."

John would climb and rest, climb and rest. Every time he would climb, he would climb a little less, and every time he would rest, he would rest a little longer. Finally I left Cynde and Beth with him and quickly climbed out to get help.

At the top of the gorge, the bus awaited and everyone else was ready to go. I got the driver of the bus to go back down with me. I don't remember his name, but he was a huge man and very strong. It was a lucky thing, too; although John's not that tall, he's built like a fireplug. Between the two of us we carried him out the rest of the way.

He looked terrible, and I thought for sure he was having a stroke, but somebody offered him a soft drink and in ten minutes he looked and felt much better. We boarded the bus, went back to our beautiful hotel, showered and changed, and in two hours John was sitting in the hotel casino drinking a beer, as if nothing had happened. He was deeply apologetic, though, for what had happened, and he was a little dumbfounded, as he had never suffered anything like that before.

An hour later we prepared to get on the train bound for Pretoria—Rovos Rail. Rovos Rail bills itself as the most luxurious train in the world, and the PR material, for once, is probably correct. Each suite on the train consists of a beautiful bedroom and private seating area and includes a private bathroom for each couple. The interior features fine wood panels and Edwardian period furnishings, and the dining car has seven beautifully carved pillars and arches, dividing the dining tables into intimate seating areas. The last car is the bar car. There are large windows along the sides, and the back wall is glass, fronting on a beautiful little outdoor porch. We boarded the train on a red carpet, in front of a table laden with hors d'oeuvres and glasses of champagne with fresh-squeezed orange juice. We had gone from life-threatening excitement to the lap of luxury.

You have to wear a jacket on the train at dinner, and the food is superb. My wife and I joined Cynde and John in the elegant dining car, and I was surprised to find that John had never felt better. But I was also surprised that they considered that I had, in their words, saved their lives. "Really, it was nothing like that at all," I protested. "I just pulled Cynde in the boat and pushed you up the gorge." But they were insistent.

"We've realized," said John, "that life is shorter than we ever dreamed, and more important than we ever gave it credit. We want to do something for the zoo to help you conserve wild places, and we want to do it right now. We learned today that you can never count on a tomorrow."

With that, John took out a pen and, on a Rovos Rail napkin, wrote out a pledge for $250,000 to help our zoo start a rhino conservation program. Cynde and John signed the pledge, and I put the napkin safely away in our luggage. I was glad that they had come to understand why what the zoo does is so important and thrilled that they had decided to support us in such a generous way. It's too damn bad that we had to nearly kill both of them in the process, though!

13

Skukuza

September 30, 2000, found us flying back to South Africa. This time, our ultimate destination was South Africa's Kruger National Park, where we were slated to track and capture five white rhinos for export back to the States, with three of them destined to live in Indianapolis.

Our group included Polly Hix, who had quietly become one of the best friends of our zoo in its entire history, John and Cynde Barnes, who had agreed to pay for the rhinos after I nearly killed them on our previous trip to southern Africa, Debbie Olson, who had by then become the zoo's director of conservation science, my wife and me, and Myrta Pulliam.

Myrta was a trustee of the zoo, and her family had owned the *Indianapolis Star,* along with several other newspapers, for three generations. She grew up working in the newspaper business and was a member of the newspaper's Pulitzer Prize–winning investigative team. She had just been appointed as director of special projects for the newspaper, following a protracted period during which she had worked with other members of the newspaper's board to sell the paper to Gannett. The sale of a family-owned paper to an out-of-town firm had caused a good deal of hand-wringing among the employees and in the community in general, but not much changed as a consequence, except that the paper got even better.

On the other hand, a lot changed for Myrta. Prior to the sale of the *Indianapolis Star,* Myrta lived exclusively off her salary at the newspaper. Granted, she happened to own a fair number of shares in the paper, but those shares didn't really have much of an established value as they were so rarely offered for sale. With the sale to Gannett, Myrta's shares were converted to stock that did have value—probably quite a bit of value.

Myrta is a delight. She is brilliant, quick-witted, and opinionated. She drinks scotch and swears when she plays tennis. She is cute, but her clothes have always been sophisticated, even before she could really afford to wear anything she wanted. As a consequence, she doesn't come across as cute, but more as worldly. I think she tends to intimidate men, even though she never sets out to do that. She simply thinks of herself as the absolute equal of any man and, as it turns out, most men find that a little scary, no matter what they say to the contrary.

Polly is also delightful to be around. At the time of our trip, she had been a sergeant with the Marion County sheriff's department in Indianapolis. She lived very modestly on her salary, but she actually was fairly well-to-do, having inherited a substantial amount of money from her father, a man whom she adored. The vast majority of the earnings from her inheritance she simply gave away, and since she loves animals, travel, and the zoo, we were the fortunate recipients of her generosity. I have always thought that she is one of the few really true philanthropists. She gives far, far more away than she ever spends on herself. In fact, her only vice is travel, which worked out perfectly for the zoo; she was always up for an adventure.

Our group was scheduled for a twenty-one-day trip, much longer than the usual fourteen to sixteen days that we allotted for a regular safari. In addition to being longer, the trip was unusual in that we intended to visit Botswana, for regular game viewing, and Namibia, in an attempt to see the rare desert elephants, before actually going out to capture animals. Finally, it was unusual in that we had our own plane and pilot for the duration of the trip, making travel far easier than usual. We could pretty much come and go as we pleased. Of course, this trip was considerably more expensive than our usual trips, but it also promised us an experience not unlike being in the middle of a National Geographic film. As it turned out, however, it was more like starring in the John Wayne film *Hatari*.

The trip started out quietly enough. Our first two stops were in the Okavango Delta at Mombo and Jao. Both of these camps abut the Moremi Reserve, and both are architectural gems. Accommodating sixteen people each, the camps are built on raised platforms over a carpet of palms. Each cabin is luxurious, and the main dining areas and adjoining elevated walkways are built from native materials. The food is simply splendid, and I think first-time visitors are always surprised. Here you are in the middle of nowhere, and the chefs are turning out elegant five-course meals that rival the food at the best restaurants in the States. Of course, part of the attraction is that they're serving food that you wouldn't often see in America—grilled oryx, zebra steak, and so on. But it is still amazing that they can come up with such marvelous soups, fresh salads, interesting vegetable dishes, and superb desserts without the sophisticated kitchen that you'd find in an American hotel or restaurant.

My first clue that this trip would be an adventure came on our second day. Riding on the third bench of our Land Rover, my wife noticed that a snake had found its way into our vehicle and was twining its way up the rail we used to brace ourselves against the bumpy terrain. It was coiled around the brace and was staring intently at John's ear, while he bumped along in the seat in front of us.

Mel said, "John, move." She said it quietly but deliberately. "There's a snake about three inches from your head."

John was nonplussed. Apparently, having escaped death on our last trip, little things like snakes had stopped bothering him. His wife, on the other hand, provided some pretty impressive vocal pyrotechnics. Her outburst caused the driver to stop, and we all got out to chase the snake out of the vehicle. Patrick was, as always, our guide, and he reacted with his usual mix of grace, wonder, and good humor. "Bloody amazing. Never seen that before. Great fun, eh?"

I was glad that we went to Mambo and Jao first, because the accommodations at our next stop in Damaraland were not nearly as plush, and this part of Namibia is not really noted for its extensive game viewing. On the other hand, the camp is fascinating. The description of Damaraland in Park East's travel itinerary actually comes pretty close to doing the place justice. The brochure talks about the brooding mass of the Brandberg Mountains, some sixty miles to the south, which provides a stunning visual backdrop for the arid plains that stretch out below the camp. Early morning mists, generated by the icy Atlantic and the warm land mass along the Skeleton Coast, drift inland along the river's channel, providing sustenance to varied life forms. The river flows only once or twice during the short rainy season, seldom breaking through the dunes to the ocean. Wildlife is not concentrated, and the natural laws of food and water availability dictate the movement and cycles of the elephant, black rhinoceros, oryx, kudu, springbok, and other species that have come to terms with life in a desert environment.

The camp itself is a series of smaller tents, each with two beds and a shower. The camp was developed after years of work with the local residents, called the Riemvasmaker people, and this project has become a sort of template for the creation of camps in other parts of Namibia. The idea is to create community tourism projects, where local people participate in planning and running the camp. This provides a strong incentive for preserving the wildlife of the entire region. The Damaraland camp was developed in cooperation with the Integrated Rural Development and Nature Conservation (IRDNC), World Wildlife Fund, and other organizations, with the ultimate goal of creating a sustainable economic asset. The local people get a regular income, receive training, and ultimately own the camp. More importantly, they collect bed levies—the equivalent of our hotel tax—from everyone who stays overnight at the camp. These funds go into a community treasury, and the money is used to build schools, clinics, community halls, and wells. But the results for wildlife are far more impressive. In the first few years after the camp was created, several species that were once fairly rare in the region began to make an astounding comeback. For example, the 1982 springbok population was only 1,000; by the time of our visit in 2000, the number had jumped to well over 200,000. Almost as impressive, given their slow rate of reproduction, was the change in the population size of desert elephants, which had grown from 280 to 560.

In Damaraland we tracked and found the desert elephants. They are still exceedingly rare but are making a slow comeback, thanks largely to ecotourism. Because they had been persecuted for so long, they are reclusive and very aggressive. The group we followed was relaxed and carefree, but three other elephants that were belligerent and aggressive soon joined them. They threw out their chests and flapped their ears in a clear warning that we shouldn't try to approach. Patrick later told us that these were the same three elephants that had killed two American tourists a few months earlier. They had, unwisely, gotten out of their vehicle to take photos, and one was trampled to death before he could even regain the comparative safety of their 4x4.

While Damaraland was fascinating, our group was ready for a break when we arrived in the town of Swakopmund. We stayed in a four-star hotel that was originally the train station; it was within two blocks of the beach and a town full of wonderful shops and restaurants. The whole place feels like a cross between a German village and a Mediterranean resort. It's filled with African antique shops, jewelry stores, a tannery that sells beautiful hides of zebra, oryx, lion, and almost any other animal, and there is a marvelous handicraft area that specializes in the creation of heavy weavings that are used for rugs or wall hangings.

From Swakopmund, we were off to Sossusvlei, where we scaled some of the world's highest sand dunes. It is a place of captivating beauty. Mountains abut plains of gravel. Undulating seas of sand spread near rugged canyons with towering walls of volcanic rock. Mountain ranges shimmer in the distance, and sand dunes the height of the Eiffel Tower seem to change color by the instant as the sun rises over them.

From Sossusvlei we moved on to the Skeleton Coast, to one of the most unusual camps in Africa. Our hosts at the camp were Dr. Ian McCallum and his wife, Sharon. He is a former psychiatrist and a world-famous rugby player, an odd combination. On the other hand, you probably need a psychiatrist to play rugby. The attraction in the Skeleton Coast, at least for me, was a chance to meet the Himba people, a group of seminomadic herders who manage to eke out a living from the harsh desert. They live in little huts made from hides and dig wells in the dry riverbeds for their cattle and goats.

The women wear a small hide loincloth and smear their bodies with a mixture of ochre and fat. Their skin is a rich red hue, and their faces are smooth and beautiful despite the almost constant exposure to the relentless sun. An enormous conch shell hangs between the breasts of the married women, and their hair is plaited in long braids that are also smeared with the red-hued fat.

We visited them a week after the camp crew had decided that the Himba should visit the lodge, so they could see where the strange white people who popped into their lives every couple of weeks came from. The crew was delighted by the experience and, to hear them tell it, so were the Himba. First, they got to ride in a car. Obviously they thought the car was alive. How else could it carry them? The camp itself held the real mysteries, though. They had never seen ice before. They had never seen water come from a pipe to their hands. They had never seen a bed or an electric fan. In fact, they had never heard of electricity, much less seen any of the fantastic devices that electricity can power. The concept of a radio, for example, was absolutely foreign, and so on. By the end of day, they were beside themselves with new discoveries. I'm not sure if it was the right thing to do, but I do wish we had been there one week earlier so we could see their response to all the modern wonders.

There is another reason I wish we had been there a week earlier: we could have avoided the first of the two potential disasters that befell us on the trip.

The first occurred at the Skeleton Coast camp that had seen the Himba visit a week prior. The camp consisted of six tents situated around a main tent that served as a dining hall, bar, and kitchen. The main tent had a canvas roof, with sides of canvas and wood. It had glass windows and screens. The screens were an absolute requirement during the day for ventilation, but the windows were critical for the night, when temperatures dropped into the fifties. The lodge was built on a hill with the kitchen and storeroom located below it. All the water and other supplies had to be flown in, and all the cooking was done with propane stoves.

On our last night in the camp, it was warm enough to eat with the windows open. Unbeknownst to us, the staff had changed propane tanks during dinner, and the new tank had not been properly connected. No one smelled gas, probably because of the open windows. Around two o'clock in the morning, my wife woke me up, saying she thought she had heard an explosion. "Go out and look," she said. "I swear I heard something."

I walked out of our tent and stood on the deck, which was pointed away from the lodge and toward a lovely, low hill. The hillside glowed with what I thought was the beautiful, orange rays of the rising sun. "It's nothing," I told my wife, "but come out and see the sunrise—it's just beautiful." I gazed at the gorgeous colors dancing on the hillside, half asleep. I wasn't the least bit angry at having been awoken from a sound sleep—the spectacular sunrise made up for it.

My wife was not half asleep. In fact, she was wide awake and possessed enough presence of mind to actually look in the other direction, toward where the sun would rise in a couple of hours. "That's not the sunrise," she snapped, "the lodge is on fire." "Grab our stuff before our tent catches fire, too!" Indeed, the sparks were rising sixty or seventy feet in the air, and as dry as everything is in the desert, the first spark that came down on our tent would likely send it up in flames more or less instantly.

We grabbed our belongings and heaved them over the porch and out on the sand below, but we really didn't have to worry. By then the camp was in an uproar; fortunately there was no wind that night, and no other tents caught fire. The lodge, however, burned to the ground. Nothing was left by morning but smoke and ashes. Sadly, the staff had kept all of their money and other valuables in the lodge. All of the food and medical supplies were destroyed, and most of the camp water had been used, not so much to fight the fire, but to keep it from spreading. Wilderness Safaris had our plane to us by ten o'clock that morning with some emergency rations and, since we were scheduled to leave that day anyway, the plane took us out. We left all our medical supplies behind, plus anything else we had that we thought might be useful until the camp could be completely restocked. We later heard that they had the camp rebuilt and open again in less than two weeks.

Our second near disaster came in Skukuza, in the southern part of the Kruger. The Kruger is one of the great parks of Africa. It is the largest game preserve in South Africa, covering about twenty thousand square kilometers. It also has the greatest concentration of species of any park in Africa, with 147 species of mammals, 505 species of birds, 109 different reptiles, and 49 different fish species in the six rivers that run through the immense preserve.

Our accommodations were in the regular tourist bungalows. They were small, round, brick buildings known as rondavels, each with a bedroom, bathroom, and kitchenette. They were well constructed and well maintained, with each cluster of about twenty buildings backing on to an open lawn area. Neighbors congregated outside in the evening, barbecuing on the outdoor grills and talking about their daily game drives.

After our arrival we took a short game drive, and then we all went back to our rooms to rest up for a longer afternoon game run. Myrta was sharing a cabin with Debbie, and Myrta as usual was reading a book. After a while, she got up and, putting her feet on the ground, noticed a fairly large snake coming out from under the bed.

Now I usually tell people that, if they should see a snake in their room, they should stand on the bed and scream. I say it as a joke, and the punch line is "It's amazing how many people know to do that without me having to tell them." Of course, I never dreamed it would happen, but for some reason people are always worried about snakes under beds. Myrta, perhaps remembering my advice or, more likely, acting on instinct, stood on the bed. She didn't scream, but she did draw Debbie's attention to their uninvited guest in a fairly loud voice. Debbie got off her bed and (in her words) "very stupidly" walked over to where the snake had slithered into their luggage. Her plan was to herd it outside but, as she approached the pile of luggage, a little voice in the back of her mind warned her against touching the snake.

Fortunately, our guide was walking past their door. Debbie called him in to see the snake. "Oh," he said, "Mozambique spitting cobra. Very dangerous." He got around behind the snake and gently shooed it toward the door. Once the snake was outside, Debbie and I joined him as he herded it toward a plastic trash can. "Careful," he kept repeating, "they can spit three feet." That didn't really sound right to me, but since I had a stick that was a good three feet long, I didn't worry about it too much. I should have. Our guide had confused feet and meters—he really meant nine feet. Anyway, we got the snake into the trash can and got the lid on the top. "We'll just take it out with us on our game run this evening and let it go."

Myrta went back to her book, and I didn't think much about the whole thing until it was time to go out on the game run. Our guide got into the Land Rover, placed a paper sack on the seat between us, and cheerfully announced that it was time to go. "Johan, where's the snake?" I asked, surprised that he would forget it. "Oh," he replied, "it's in the sack." My face must have transformed itself into a mask of sheer horror. The sack was made of paper and was sitting about six inches from my leg. I had visions of an angry cobra spitting his way out of a paper bag and taking out his displeasure on my thigh. "What?" I shrieked, "are you crazy?" "Don't worry," he said placatingly, "I double-bagged it."

Cobras aside, the real reason we were in Skukuza was to capture white rhinos for the Indianapolis Zoo. Rhinos are still highly endangered, but the white rhino, in particular, has staged an amazing comeback. There were only about one hundred white rhinos left alive at one point in the last century. Now, thanks to zoos and a host of other wildlife conservation organizations, the number of white rhinos has grown to some eleven thousand. Of the five species of rhino, the Javan rhino is still in critical condition. As of this writing, there are about sixty left. That does not mean that their fate is sealed, but the prognosis certainly is not good.

There are five subspecies of rhinos altogether, and they are all found either in Africa or Asia. The Asian species include the Indian rhino, which has skin that looks just like armor plates, and the much rarer Sumatran and Javan rhinos. Sumatrans are not quite as endangered as Javans but are still in serious trouble. There are now about three hundred left alive. In Africa there are two species, the black rhino and the white rhino. Their names don't really refer so much to their color, which is the same. The white rhino gets its name from the German word for "wide," a reference to its wide mouth, which helps it to eat enormous swaths of grass. In fact, they are often described as a giant lawn mower on legs. The black rhino, on the other hand, has a pointed lip that makes it an efficient browser of trees and branches.

All rhinos have elongated heads, small eyes on either side of the head, and upright, active ears. Their sense of sight is poor, and because of the location of their eyes, they have difficulty seeing straight ahead and must actually swivel their heads from side to side to see what's in front of them. Their senses of smell and hearing, on the other hand, are acute. Perhaps it is their inability to see danger when it's literally right in front of them that gives them the reputation for being, well, dimwits. But they're actually quite bright and very aggressive.

As a rule, rhinos are not very sociable. Mothers stay with the young while nursing, but on maturity the young leave. Female white rhinos sometimes gather together in small, ephemeral herds, but males are very territorial and aggressively work to keep other males away. They do this by marking their territory with great piles of dung and by spraying powerful streams of urine in backward jets. If neighboring males meet, one may show signs of subordination, perhaps by turning back its ears and adopting a submissive posture. If that doesn't happen, they may use their horns in a battle to the death. Males need to maintain a territory in order to breed, so they tend to start reproducing later in life, when they're big enough to grab and defend a territory of their own. Females start breeding around age five or six. Gestation is long, about sixteen months, and it is only when calves are young that they are vulnerable to predators. Rhino moms are, however, very protective of their young. A rhino is capable of killing a lion in defense of its baby.

Rhinoceroses get their name from the Greek words for "nose" (*rhino*) and "horn" (*keras*). The horn is made of a mass of keratin, the same substance that is found in our hair and nails. The horns are permanent and grow throughout the animal's life. They are formidable weapons. Think about it. An adult white rhino can weigh anywhere between five thousand and seven thousand pounds and can run at speeds of up to thirty miles per hour. Imagine all that mass and momentum concentrated behind the tip of a very sharp horn. But just as that horn can save its owner's life, it can also spell its owner's demise.

In some ways, it is shocking that the rhino population ever got so low to begin with. After all, they are among the most charismatic animals on the planet. They are four thousand pounds of armored attitude. The second largest land animal after the elephant, they resemble living dinosaurs as they move across the landscape. They fear nothing. They will charge anything. And they have only one true enemy: us.

Humans have hunted the rhino almost to extinction. Perhaps because the male can copulate with a female for an hour on end, or perhaps because of the phallic symbolism of their great horn, they have come to be regarded, at least in some parts of the world, as a symbol of male virility. Powdered rhino horn is thought to be the ultimate aphrodisiac. Their horn is also used in some Middle Eastern cultures to make a ceremonial dagger that is used to mark the occasion of a young man reaching adulthood. Perhaps people equate the phallic symbol of the horn with the knife, that most phallic of weapons. Or maybe people are just nuts. In any case, the one place on the planet where black and white rhinos thrive is in the Kruger National Park, in South Africa. They thrive here for one reason and one reason only: there are men with guns here who will shoot you if they find you someplace where you are not supposed to be. In the Kruger, the war on poachers is, in fact, a shooting war.

The Kruger is named for its founder, "Oom" Paul Kruger, affectionately called Uncle Paul. He was obsessed by hunting in his youth, but by middle age he had grown into an ardent conservationist with a deep concern for the rapidly disappearing animals of his nation. In 1884, barely a year after he became president of the Transvaal Republic, he put forth legislation to create a wildlife sanctuary.

It seems like an obvious thing to do now, but at the time it was a revolutionary proposal. The Volksraad (the congress of the Transvaal) was composed of dour, pragmatic farmers. To quote from a wonderful book, *The Wildlife Parks of Africa,* the Volksraad:

> believed in the Bible, in the family, in money, and in agriculture. If they thought of Africa's wildlife at all, they thought of it as vermin. Kruger's proposal was shocking, offensive, and heretical. Shocking because it came from one of their own kind; offensive in that he also happened to be their chosen leader and so, of all people, should have known better; and heretical because it was a clear repudiation of God's word. Kruger was not only suggesting the lowest forms of life should be protected, he was also arguing for precious land to be put aside for them to infest. It was nothing short of defying the Almighty and denying food to His own children.

Not surprisingly, it would take ten full years before Kruger could put aside any land at all, and it was not until 1898 that Kruger could marshal the political clout to create the Sabie Reserve, the land that would eventually become a national park bearing his name. By 1902 the British, victorious in the Anglo-Boer War, had proclaimed the Sabie a true reserve and gave the job of game warden for the reserve to Colonel James Stevenson-Hamilton. He thought he'd have the job for about eighteen months, until he could return to his regiment and career. He wound up working in the Kruger until 1946, turning the reserve into one of the wonders of Africa and one of the wonders of the world.

By the 1980s, the Kruger had a surplus of rhinos. There is a common myth in the minds of many people that animals are endangered because they don't reproduce, either in zoos or in the wild. In fact, nothing could be further from the truth. By and large, animals reproduce very well. If they are safe from the depredations of people, their population levels will rebound to the maximum level that their habitat can support (this level is called the "carrying capacity"). Around 1988, the wardens of the park began to round up rhino for relocation to other parks and to zoos; by 2000, they had relocated some six hundred animals.

The managers of the park were excited about capturing rhinos for Indianapolis for three reasons. First, they realized that it makes sense to have a stable population of rhinos outside of Africa as a sort of insurance policy, a hedge against extinction. A widespread disease could quickly reverse a century of rhino conservation. They were particularly enthusiastic about Indianapolis because of what we had done with elephants. Just as there was little known about the hormonal basis of elephant reproductive physiology, there was (and still is) not much known about hormones and reproduction in rhino. Indianapolis had the staff and expertise to unravel those mysteries. Finally, we were willing to pay the costs of collecting the rhino, plus we were willing to make a pretty sizable donation to the park, helping to further their overall conservation efforts.

And so, one morning at about 3:30, we found ourselves traveling out to the veld with a capture team in search of three white rhinos to bring back to the Indianapolis Zoo. Dr. Douw Grobler, then the Kruger's chief veterinarian, led the team. The procedure for capturing rhino is a fascinating mix of high and low technology, with the team consisting of a ground crew with two huge flatbed trucks. The trucks are equipped with hydraulic lifts, and each carries three specially designed rhino crates. Several Land Rovers, a few tons of heavy equipment, and a whole lot of rope make up the rest of the supplies.

When the team is assembled on the ground, it's met by a four-seat helicopter that's modified to allow a rifleman access to shoot out the open door from the back seat. The helicopter lifts off and begins to fly a grid pattern near the vicinity of the ground team. The spotters in the helicopter (in this case, the pilot and me) search the ground for rhinos. Even one hundred feet in the air, an expert (in this case, not me) can tell the difference between a male and female and can pretty much gauge the age of an animal, primarily from looking at the size of its horns, as well as the size and shape of the rhino's neck and body.

After about twenty minutes in the air, we spotted our first candidate for an all-expenses paid, one-way trip to Indianapolis. She appeared to be a female somewhere between three and five years of age. Dr. Grobler, riding in the back seat of the copter, prepared an anesthetic cocktail known as M99, a potent blend of a morphine derivative and a tranquilizer known as azaperone. He loaded the M99 into a special syringe, which was then inserted into a rifle. Opening the back door, Dr. Grobler calmly shot our rhino in the butt.

A similar procedure is used to dart elephants and, at least with elephants, a good helicopter pilot can herd the animal directly toward the ground team. This sometimes works with rhino, but usually they start to run in whatever direction they're pointed despite the best efforts of the pilot. We were lucky, though, and our rhino took off directly toward the ground crew. The trick is, it takes about ten minutes for the drugs to kick in. Ideally, they work before the rhino can get close enough to the ground crew to do any real damage.

Eventually our rhino went down, and our pilot set us down twenty feet from the sleeping animal. The capture team arrived a few minutes later, and we all set to work. The first order of business was to blindfold our rhino. This keeps them calmer as they awake from the drug. Vital statistics are recorded, blood is drawn for analysis, and the truck is positioned to load the animal.

This is where the low-tech part comes into play. Getting the rhino into the crate is mostly a matter of pushing and pulling. A rope is looped around the horn, pulled through the front of the crate, and three or four big guys take hold. At the other end, the rest of us find a convenient patch of rhino backside and get ready to push. If you've ever wondered, a rhino's butt feels much like the rest of a rhino—muscle covered by sandpaper.

Next came the tricky part. Not even ten of us could push a sleeping rhino into a crate. We actually needed the rhino to cooperate, at least a little, and for that the rhino has to be awake, so Dr. Grobler prepared a "partial antidote" to the anesthesia. When everyone was in place, the drug was administered and the rhino snapped right out of it. As she struggled to her feet, we pushed and pulled her directly into the crate and dropped the door. If anything goes wrong at this point, it's pretty much every man for himself (or every woman for herself, since Cynde, Myrta, Polly, Debbie, and my wife were pushing and pulling along with the rest of us). Before giving the shot, it's wise for everyone to figure out the shortest path to the relative safety of the flatbed.

Once the rhino was in the crate, the hydraulic lift hoisted her onto the truck, and we left for the boma, or capture pen. The boma is made of stout tree trunks that have been driven into the ground to make a secure holding area, and the captured rhinos are released one by one into the pen. Rhinos have definite likes and dislikes, and some simply don't want to be kept in a pen and fed three square meals a day. Others may snort a few times, paw the ground once or twice, and then placidly start to have lunch. If a rhino doesn't seem to be liking its new home, after a few days the keepers simply open the gate and allow the rhino to trot back into the wild. If the rhino does seem amenable to its new digs, it's kept there long enough to be sure that it's not carrying any diseases and, more importantly, long enough to get used to being around people. Nine out of ten rhinos do just fine.

After our first capture or two, we really began to get the hang of it. All of us got at least one ride in the helicopter. By the time we got to the fifth rhino, we were swaggering like John Wayne. Of course, we weren't riding on the hood of a 4x4 and lassoing them, so the swagger wasn't quite as cocky. We were also pretty much ready to go home. Even rhino wrangling has its limits, and I think, after all the excitement of the trip, most of us were ready to return to a more normal existence.

It would be several months, however, before our three rhinos were ready to travel, and when they were, they did not go directly to Indianapolis. Instead they went to the Fort Worth Zoo because the Indianapolis Zoo wasn't quite ready to house them. They were slated to occupy the old elephant exhibit, which wasn't large enough to house the existing elephant herd plus the addition of two new babies. This was especially true because one of the babies was a bull, and bulls, of course, require a special facility. So while a new elephant barn was under construction, Fort Worth agreed to keep our rhinos.

The arrangement was fine with Debbie. Two years earlier she had met and married the director of the Fort Worth Zoo, Michael Fouraker. Even though she was still employed by the Indianapolis Zoo on a full-time basis, she found it fairly easy to work out of Texas. Since my arrival at the zoo in 1992, she had been spending most of her time working on international elephant conservation issues. All she really needed for that was a computer and a whopping big budget for long-distance calls. It didn't really matter to anyone where her desk was (well, it mattered to her and Michael, but it didn't much matter to anyone else).

Michael Fouraker is an internationally renowned authority on rhinos. He has spent a good deal of his life preserving rhinos from extinction and has a unique perspective on why it's important. He compares the loss of a species, any species, to losing the rivet on a wing of a plane. "Who cares," he asks, "if you lose a few rivets?" "On the other hand, if you lose enough of them, eventually the wing will fall off." "Rhinos," he continues, "are what we call umbrella species. They have a major impact on the habitat in which they live. This makes them pretty important rivets. Lose the rhino, and you may lose the one rivet that causes the plane—or the habitat—to go down."

Given their long years of experience in dealing with rhinos, the staff at Fort Worth has become expert in training them, so in addition to caring for our animals, they initiated their training program. The first part of the training involved getting the animals to home in on a target—in this case, a ball on the end of a stick. When a trainer gives the command, the rhino is supposed to go up to the stick and touch it with its head. When the rhino does this, it gets a treat. Eventually the rhinos learned to come and stand still, and at some point the trainers finally get rid of the stick and target entirely. By the end of the training period, the rhinos can do all sorts of things, including walking into their shipping crates on command. This makes moving them a whole lot easier.

By May 2003, the rhinos had a new home ready in Indianapolis; they are now participants in some very important research. Thanks to the training in Fort Worth, they will come to the edge of their enclosure and allow the staff to draw blood—in return, of course, for something good to eat. Eventually the research that they participate in will help us to better understand how they reproduce and what we can do to help them. We're not exactly sure what we'll discover or how it will help, but that's in the nature of scientific research. Sometimes scientific discoveries have completely unanticipated effects. My favorite example, at least as it relates to rhinos, is the discovery of the drug Viagra.

In a weird twist of pharmaceutical karma, it may be that Viagra will ultimately save these great beasts. Once people find that there is an aphrodisiac that actually works, they may begin to leave the rhino alone. It is ironic to think that, with all the effort of so many people, with all the time and money invested in just bringing rhinos to Indianapolis, much less all the time and money invested in preserving and protecting these wonderful animals in the wild, the person who makes the greatest contribution to their preservation might turn out to be a scientist working in a laboratory at the Pfizer Corporation!

14

The Panda Wars

The panda wars were civil wars in the truest sense of the phrase. They were civil in the literal sense, in that no one got shot at (although, at times, I'm sure the thought crossed people's minds), and civil in the more common sense, in that they pitted conservation organizations that were normally the closest of friends and allies against one another. The panda wars reached their height in 1992 and turned on the desire of zoos, in particular the Toledo Zoo and the Columbus Zoo, to exhibit pandas against the desire of our national zoo association (the American Zoo and Aquarium Association—the AZA) and World Wildlife Fund USA (WWF) to restrict their exhibition or to ban short-term exhibition (say, three to six months) entirely.

The strangest thing, at least from my point of view, about the panda wars is that no one ever actually sat down and told me the story. I had to piece it together for myself in dribs and drabs, although, had I but known, most of the story was already published in George Schaller's book, *The Last Panda*. Someone *should* have told me the story long ago, if only because three of the principal actors in the saga were people I knew well—Jerry Borin from the Columbus Zoo and Bill Dennler from the Toledo Zoo, who were two of the zoo directors who accompanied me on my first trip to Africa, and Rich Block, who at the time of the panda wars in the early 1990s worked for WWF, and later worked with me in Indianapolis for several years before becoming the director of the Santa Barbara Zoo. But not one of them ever brought up the subject of the panda wars in all of the time I've known them. And that's a shame. It's a shame because the story of this conflict illuminates, perhaps better than any other story I can tell, the politics of international and national conservation and the economics of running a zoo while benefiting animals in the wild.

We'll begin the story in St. Louis in 1939, just prior to the outbreak of World War II. The decade from 1936 to 1946 saw fourteen pandas leave China for Western zoos. Very little was known about pandas at that time. They live in very rugged mountains, primarily in the western area of Szechuan, China. Historically, they were widespread in eastern and southern China, but today there are only about one thousand pandas left in the wild, scattered among six isolated areas of three provinces in southwestern China. Secretive and solitary, they live in the dense bamboo forests at elevations of 4,000 to 9,000 feet. They were first known to the Western world from a pelt sent to the Paris museum in 1869 by Father Armand David, a French missionary and scientist.

Despite the fact that they look like bears, scientists felt they were probably more closely related to raccoons, at least until recently, when genetic testing revealed that they were indeed more akin to bears. Largely (although not exclusively) vegetarians, they feed primarily on the leaves and shoots of the bamboo, which grows so dense that few other mammals can live, or find nourishment, in the thickets. The bamboo forests thus feed and protect an otherwise largely defenseless and conspicuously marked animal. Really, pandas were designed by nature to be carnivores, and their digestive system isn't very good at breaking down the cellulose in bamboo. As a consequence, they need to eat up to forty pounds of bamboo every day.

The biggest problem with specializing in a diet of bamboo, however, has to do more with the life cycle of the bamboo plant than it does with the plant's nutritional value. Bamboo flowers regularly, anywhere from every ten years up to every one hundred years, depending on the species. Once it flowers, the plants die, all at the same time. This means that an entire mountainside of bamboo might die all at once. Of course, the plants regenerate themselves from the seeds that developed when the plants flowered, but it can take twenty years before the new plants grow large enough to support pandas again.

Long ago, the pandas could just migrate to another place with a different species of bamboo, but even by the 1930s that was no longer possible in much of China. A significant amount of their habitat had been cleared for agriculture, firewood, or timber for construction. Roads and towns have sprung up, and pandas are now confined to the remaining fifteen panda reserves located in the six major panda habitat areas. Even there, they are not completely safe. Pandas are poached for their skins, which have been sold for as much as two hundred thousand dollars in Japan and, even more often, they are caught in the snares that poachers set for deer and other small game.

With the first pandas to leave China in 1936 came an international frenzy of interest in this already rare but truly delightful creature. An exotic animal hunter named Floyd Tangier Smith had traveled to China to attempt to collect animals from the wild in the 1930s. He subsidized Chinese hunters, who would, in turn, pay farmers, medicinal herb gatherers, loggers, and anyone else who ventured into the bamboo forests for information on where pandas could be found. While we don't know exactly how they were trapped, it's a fair guess that more pandas died than made it successfully into Smith's hands.

By 1936, Smith had captured five pandas. He carried them in trailers to Tungshan, on the eastern coast of China, and then on to Hong Kong and later to Europe. Three of the pandas were originally destined for the London Zoo, and the remaining two were to be shipped to our zoo in St. Louis. As a footnote, Smith died by the time his pandas finally reached America, probably as a direct result of health problems he developed while in China.

Smith had enlisted a well-known animal dealer named Heinz Ruhe to sell the animals; George Vierheller, then director of the Saint Louis Zoo, had dealt with Ruhe on many occasions. In 1939, Ruhe called Vierheller to say that one and maybe two animals were finally available for sale, and Vierheller, along with one of the zoo's trustees, rushed to New York to close the deal. On their arrival, they learned that one of the pandas had died but that the other, whom Ruhe had named Happy, was in good condition and could be had for seven thousand dollars. Vierheller described the events in an unpublished manuscript that I found in our archives, and it's worth letting him tell this part of the story in his own words.

> Heinz was a good friend of mine and I always enjoyed price haggling with him. Whenever we were together we were very much like two vociferous traders at a typical Far Eastern mart. As was our custom, shortly after I reached New York, we were bickering in a friendly and spirited fashion between drinks at the Oak Room of the Plaza Hotel. Heinz was in good form that day and seemed to be outarguing me. In

fact, I had just about given up hope of getting a price reduction when coincidence suddenly worked in my favor. I think I had just challenged him with my final point. I was accusing him of offering me a tired animal—one that had been on exhibition tour in Europe before being brought here, when Caesar, the maitre d' at that time, came up to me and said that my secretary was calling from St. Louis. When I talked to her I learned that a Mr. and Mrs. Schultz had cabled from Lewing, China, and had offered to donate a young panda to us if we would agree to pay the transportation costs, approximately $2,000. The couple's home had originally been in Ferguson, a suburb of ours, and they naturally wanted to give their home section advantage of their find. Well, when I returned to the Oak Room, I'm afraid that I wore a very self-satisfied smug grin. I told Heinz that he could peddle his animal somewhere else—that I was no longer interested in it because someone was giving me one! He looked at me in amazed disbelief, sputtered a few choice remarks, and then, when he realized the unexpected had actually happened, agreed to sell Happy to me for $5,000. Naturally, we accepted.

George Vierheller. SAINT LOUIS ZOO

Vierheller brought Happy from New York by train. The animal was, to his thinking, clearly uncomfortable, so in Pittsburgh they opened the door to the baggage car and put ice cakes on the floor. On his arrival in St. Louis he went almost immediately on display and, by all accounts, seemed to like his new surroundings, especially his little plunge pool. He was fed bamboo harvested from the Missouri Botanical Garden or flown up from Florida, plus a mix of apples, carrots, celery, spinach, and lettuce, supplemented with cod liver oil, honey, oats, pablum, baby soup, and evaporated milk. According to Vierheller, Happy liked the recipe, as witnessed by the fact that he licked his platter clean, pounded the bottom of his overturned plate for more, and steadily gained weight and conditioning.

If Happy was happy in St. Louis, it could be equally said that St. Louis was happy with Happy. On the first day he went on exhibit, a Sunday in June, the zoo drew a record-breaking forty thousand visitors. Fifteen thousand people a day, on average, came to view him, even though, as the summer arrived, Happy went out in his exhibit yard less and less due to the heat. But, says Vierheller, the crowds waited patiently to see him and, whenever he made a brief appearance in the yard, they cheered.

Meanwhile, Vierheller was engrossed in getting his second animal out of China. The Chinese government had put restrictions on the export of pandas just after Smith's animals left the country, but Vierheller argued that he made transportation arrangements before the new regulations went into effect, and so he did manage to get an export permit. The panda, named Pao Pei, or Precious Jewel, was to be shipped to Vancouver and was due to arrive in September, some four months after Happy's debut. Travel arrangements were complicated, though, by an unfortunate incident. In the spring of 1939, a panda being brought to the Bronx Zoo by another collector broke loose in the airplane and caused what Vierheller termed a "minor panic." Naturally, commercial carriers refused to carry any more pandas. Vierheller didn't want another long train ride, but it seemed that he would have no choice. To make matters even more complicated, Canada was, by September 1939, officially at war. For security reasons, Vierheller could not be told when Pao Pei's ship would even come in. If Vierheller wanted to fly his panda to America, he would need to charter a plane, something he didn't feel the zoo could afford. Then, as he says, "suddenly an idea clicked." I'll let him pick up the narrative again.

Frank Phillips of Phillips Petroleum was a good friend of mine and I decided to try to inveigle him into lending us his private plane to transport the panda. I figured that the trip could work into a good ad for the oil company and, at the same time, would help us out of our dilemma. I judged that the panda would probably weigh about 66 pounds upon arrival and I knew that we would undoubtedly stop at several cities important to the Phillips 66 Co. So, with those key points in mind, I called Frank in Bartlesville one morning shortly after the first of September and asked him for the use of his plane. The wire was dead for a while after my request—then he said, "George, you're drunk! And I'm in a meeting! Good-bye!" The refusal seemed emphatic, but I wasn't ready to give up. Knowing Frank, I decided to wait until the last minute, state my deductions to his secretary, and not give him a chance to talk back. Mrs. Vierheller and I planned to leave on September 6th, so, on the 5th I called Frank's secretary in New York and said, "Get in touch with Mr. Phillips. Tell him to take two or three seats out of his plane and build a cage to occupy the space. Then have the plane pick up Mr. Lemp [the Director of the Seattle Zoo] in St. Louis and meet me in Seattle not later than the 10th. We can arrange stop overs in Denver and Kansas City where you have big subsidiaries. I think it can be a good deal for both of us. But don't call me back. I'm leaving tomorrow morning. Tell him that I'll expect to see that plane."

With that, Vierheller was off to Canada. He knew that Pao Pei was coming in on the *Empress of Russia,* but didn't know precisely when. Neither the American consul nor the Canadian Steamboat Company would give up any information on the whereabouts of the *Empress.* Discouraged, Vierheller did exactly what I would have done. He went to the cocktail lounge of his Vancouver hotel and voiced his frustrations to the bartender.

As it happened, the bartender knew the pilot of one of the harbor escort ships. The pilot knew the approximate time of arrival, and Vierheller bribed the night bellboy to watch the harbor from the roof of the hotel so he could be at the dock the moment the ship came in. He still faced, however, the problem of getting on board. Again, Vierheller picks up the narrative.

I wandered around the dock aimlessly for a short while, and then I recognized the attitude and the equipment of a group of photographers. I asked them what they were after, and when I learned that they wanted pictures of the panda, I talked fast. I told them that I owned the animal and that I wanted to get on the boat to pick it up. I had always been on good terms with the press and knew their lingo. So, without any further delay, and as a temporary member of their clan, I was taken aboard to search for my pet.

Eventually, Vierheller found Pao Pei curled up in a wicker basket in one of the ship's staterooms. By this time, passengers were already getting in line for their quarantine inspection, and it looked like Vierheller would have a long wait. Again he did the logical thing. He bribed the purser, and Pao Pei was first off the ship. (The *Empress of Russia*, by the way, came to a sad end. She was destroyed by fire during her postwar refitting.)

Pao Pei went by taxi to the nearby ferry for Seattle, with Mr. and Mrs. Vierheller stewing the whole time over whether the Phillips 66 plane would be in Seattle to meet them. It was. They flew first to Denver, where a crowd of about five thousand greeted them at the airport, and then to Kansas City, where another huge throng greeted them.

Pao Pei turned out to be a female and Happy, a male. Naturally Vierheller dreamed of panda babies, but they were never more than just friends. "Pao Pei tried to be coy and amorous but Happy wasn't interested," said Vierheller. "Since a panda's age is a quality to guess, not to know, and since Happy appeared to be fully grown when he arrived, we finally decided that he had probably been older than we surmised and was, therefore, reluctant to enter into family life."

Happy and Pao Pei, the two Saint Louis Zoo pandas. SAINT LOUIS ZOO

Happy died in March 1946 of heart disease and hardening of the arteries. Pao Pei, the last panda to be publicly displayed in an American zoo until 1972, when China gave the National Zoo two pandas as a gesture of goodwill diplomacy, died in June 1952. One last time, I'll turn to Vierheller's narrative.

> We noticed that she was becoming listless and gradually losing her interest in play. She lost weight and, in spite of all our efforts, her body seemed to waste away. Finally, her hind legs became paralyzed. Her well-wishers sent in many remedies, and everyone hoped that she would recover from her malady, but there was no help for her. Her illness was finally diagnosed as a complete vitamin deficiency. Either she could not assimilate her food, or her body lacked some mineral substance that we were not able to supply. The whole city grieved when we lost her.

And so, after thirteen years, the last panda in St. Louis was gone.

I started the story of the panda wars, which took place in the late 1980s and early 1990s, by telling the story of the St. Louis pandas, because every major element of the panda wars was presaged by our early experiences. Our zoo, in all honesty, showed very little interest in conserving pandas in the wild. They were here to help us, and not the other way around; we did very little systematic research on pandas during the thirteen years they were here, even though there were clear problems in the areas of reproduction and nutrition. We were motivated primarily by a legitimate desire to draw people to the zoo, and we were not the least bit blind to the economic benefits of having them. We were more than a little bit willing to bend the rules and use whatever political clout we had to make sure they got here and got here safely. More than a couple bribes changed hands in the process. We knew they could generate tremendous press and took full advantage of the media's interest. Finally, we were fully cognizant of their ability to help major corporations from a public relations standpoint, and we took full advantage of this, too. But by the 1990s, things should have changed. I mean, if we've seen one thing in all of the stories in this book, it's that the zoos of today are far different than the zoos of fifty or even twenty years ago. The surprising thing, at least to me, is that we're so different now than the zoos of even ten years ago. Let me fast-forward to 1984 and pick up the story again.

In 1984 China presented two pandas to the Los Angeles Zoo in celebration of the Olympic Games. This was a gesture of goodwill and no fee was charged, but then the Chinese allowed the pandas, three months later, to travel to the San Francisco Zoo, this time for a hefty rental fee. Thus began the "rent-a-panda" era. Between 1984 and 1987, zoos in Los Angeles, San Francisco, New York, San Diego, Memphis, and Busch Gardens (in Florida—with the Busch Gardens pandas coming from the Bronx Zoo *without* a permit), plus zoos in Japan, Belgium, the Netherlands, Australia, and New Zealand were involved in panda rentals. Zoos in Calgary, Atlanta, Columbus, Portland, Seattle, and Toledo were bargaining to rent pandas in 1988 and beyond, with Toledo competing with Detroit's Michigan State Fair to be the first to bring pandas back to the Midwest. Pandas had become big business for the Chinese. They brought in foreign currency from the rental fees and, depending on the contract, a percentage of gift shop sales, and they brought the Chinese recognition, opportunities for foreign travel, and, I believe, great goodwill from a fascinated and appreciative foreign audience.

But it also caused some committed panda experts, in particular George Schaller, who had worked in China with the pandas since 1980 with the support of the Wildlife Conservation Society and WWF, to question the ethics of short-term exhibits. Schaller summarized the whole debate by saying:

> Zoos are cultural and educational institutions whose exhibits make people aware of rare animals and the need to protect them, and pandas are, of course, superb at raising public consciousness. However, a good case can also be made against public rentals. One can rightly argue that captive breeding in China should achieve sustained success before individuals are sent on stressful world trips during which their reproductive life is wholly disrupted. After all, China still removes pandas from the wild to augment a captive population that dies faster than it reproduces. Initially, I favored strictly regulated loans because I felt that an open-door policy with China was needed, a policy that would enable foreign zoos to cooperate with the Chinese and provide encouragement, knowledge, help, money, and, where needed, pressure to improve panda management in the wild and in captivity. But I changed my mind after observing the greed, politics, lack of cooperation, and undisciplined scramble for pandas that characterized the whole loan program.

By the time the Toledo Zoo was ready to import pandas in 1988, the panda wars had come almost to a head. George Schaller had joined with WWF to protest what he felt was a mad scramble to acquire pandas for zoos without serious regard to their welfare in the wild. AZA joined WWF, who opposed continued importation of pandas for display for five main reasons. First, WWF maintained that the loans violated the CITES agreement restricting trade on threatened and endangered species because they were "primarily commercial" in nature, meaning that the conservation benefits were, at best, poorly defined. Second, in 1988 the Chinese still hadn't approved a plan for an overall effort to preserve pandas in the wild, making it hard to set conservation priorities. Third, the money that flowed to China for rentals couldn't really be effectively accounted for by the Chinese, and WWF feared that much of the money would go for other efforts that were not related to pandas. Fourth, we had no SSP in place for pandas and consequently didn't have a good sense of which individual animals should be bred together. Finally, WWF (and AZA) felt that there was simply no coordinated plan for American zoos to cooperate on research and conservation efforts.

At the urging of Rich Block and Ed Schmitt, George Rabb, then the director of Chicago's Brookfield Zoo, convened a strategy session designed to develop a coordinated plan for pandas. Bob Wagner of the AZA, Rich Block, Chris Elliott, and George Schaller of the Wildlife Conservation Society, and several other zoo professionals met to create a strategy for a cooperative effort between the AZA and WWF. It was this meeting that would lay the foundation for an effort that would result in a court case against the USFWS over the Toledo Zoo's proposed importation.

Rich Block, who was then employed by WWF, collected most of the information that Schaller and WWF brought forth. Early on, Rich was a proponent of pandas coming to the United States, feeling that the education and awareness-building that zoos could provide would only serve to help the cause of panda conservation. Rich was particularly enthusiastic about having pandas in Toledo. He had met with Bill Dennler, the zoo's director, and felt that they had a wonderful education program planned for the pandas and that they had a great exhibit area. But evidence continued to come Rich's way indicating that, by moving so many breeding pandas out of China, there was an extremely high possibility that, biologically speaking, the loan program would do more harm than good. In fact, if everyone who wanted pandas on short-term loan got them, Rich guessed that about half of the Chinese breeding population would have been on the road with no real understanding of how this could affect future breeding! In the end, this is what motivated AZA and WWF to change their stance on panda importation.

But Toledo was dead set on bringing pandas in, so when WWF and AZA came out against their importation, Toledo sued both of them. This took place just as WWF and AZA were suing US Fish and Wildlife. One of the concerns of WWF and other conservation organizations was that the short-term loans were "primarily commercial" in nature, in that they were designed to boost attendance (and thus revenue) for the host institution. This, of course, was not the position of the host institutions. Ironically, though, the suit filed by Toledo against WWF and AZA had as its basis "interference in commerce."

Toledo's suit served to slow down the regulatory process enough that Toledo got their pandas in just under the wire. Don Reed, a Canadian who worked for WWF in Wolong (one of the larger breeding and research centers in China), was very familiar with the animals proposed for loan to Toledo. He had reviewed the USFWS permit that allowed Toledo to bring an endangered species into the United States and, upon visiting the Toledo Zoo at the request of WWF, was shocked to find that the animals on display in Toledo were not the ones specified in the permit. Toledo had unknowingly brought breeding animals into the country—precisely what WWF and AZA had feared would happen.

The upshot of the Toledo Zoo's efforts to bring in pandas was that, in 1988, AZA's giant panda task force (consisting of a group of some twenty zoos interested in pandas, including the Saint Louis Zoo, which had no interest in displaying pandas but did have a considerable amount of historical experience) issued some very strict guidelines against short-term loans. It was not enough to stop Toledo from displaying animals, but it drew a line in the sand. In 1990, the AZA adopted even tougher guidelines, including a full moratorium on the acquisition of giant pandas. The new policy stated that a "violation of the moratorium will be considered a breach of the Association's Code of Professional Ethics." (The chair of the giant panda task force was Devra Kleiman, the same woman who formed the highly effective conservation breeding program for the tamarins of Brazil.) A conservation plan, originally crafted in 1989, was submitted to the Chinese, and considerable progress was made toward outlining a research plan that detailed how American zoos could benefit pandas in the wild by studying them in zoos.

Every American zoo bought into the agreement, except for the Columbus Zoo, then headed by their popular and well-known director, Jack Hanna. Columbus had gone to extraordinary lengths to secure a panda and simply didn't want to sign on with the other zoos. They decided to bring pandas to the Columbus Zoo. A team from WWF flew to Columbus to make an eleventh-hour appeal to the zoo and to several members of the zoo's board to withdraw their loan plans. This was unsuccessful, and once again WWF positioned itself to sue the US Fish and Wildlife Service. The zoo, following Toledo's tactics, sued WWF, preventing any last-minute revoking of the permit. The AZA, in turn, decided to suspend the Columbus Zoo's accreditation and to revoke, for a year, the professional status of Mr. Hanna. This was the stiffest sanction AZA could impose. It didn't stop the Columbus Zoo from bringing in pandas, but it did mean that the zoo lost all professional standing.

I really didn't know much about any of this until 2002, when the AZA held its annual conference in Columbus, Ohio. Jack was the keynote speaker for the conference, and at one point in his talk, he told the audience of zoo professionals that he had been a proud member of AZA for his entire career—every year save one. I was sitting a long way back in the enormous ballroom where Jack was speaking, but I could see his face on giant TV monitors on either side of the stage. When he mentioned that he had been a proud member of AZA for every year of his career save one, I thought I saw him smile wryly, but perhaps that was just my imagination.

What I'm certain of, however, is that a small tear slid down his cheek when he began to talk about what was most important to him. "I've always tried," he said, "to do the best job I could on my television show and my appearances on national TV to promote wildlife and zoos." "But I never really cared," he continued, "about what the general public thought as much as I cared about what you in this audience thought. You are the ones I want to be proud of me—your opinion matters more than anyone else's." Jack lives on a big ranch in Montana for most of the year. But the ranch is not as big as his heart. I took him at his word. I don't think for a minute that Jack thought his zoo was doing anything other than the right thing.

So there it is. Zoos pitted against other zoos and against our national association, and the WWF pitted against its most important ally in the effort to help the American public understand the crisis that faced (and still faces) this marvelous animal. Zoos split internally, as was the case with the Bronx Zoo, which was asked by the city to display pandas even though one of their most important and influential field scientists, George Schaller, was against it. And, finally, a federal agency (the USFWS) that was caught between powerful competing interests and was, at best, uncertain how to resolve the many conflicts that had arisen around the issue of short-term loans.

In 1993 the USFWS exercised its authority under the Endangered Species Act and slapped a moratorium on processing any and all panda importation licenses. The war was officially over.

Now there are several zoos in the United States with pandas, and they are all doing wonderful work. The San Diego Zoo has had success in breeding pandas, the Atlanta Zoo continues to do groundbreaking research with this species, and other zoos are in line to continue this effort. American zoos that house pandas pay an annual fee of one million dollars to keep pandas on display, all of which goes to conservation. Still, there are lingering doubts. The USFWS has experienced many difficulties in figuring out the wisest ways of spending the money for conservation in China, and the pandas' future, while improving, is still bleak. Their numbers have increased from an estimate of 1,100 in 1988 to 1,600 at the end of 2004 (although there is some question whether this is a real increase or simply a more accurate census), but there is no guarantee that their future is secure.

At the same time that Columbus was locked in a struggle to bring pandas to Ohio, a completely different battle was begun by a collection of zoos, major universities, and international conservation organizations from around the world. This battle was less a civil war and more what I would call a *revolution*. The battle was to save the unique and wonderful animals of a large island off the coast of Africa—Madagascar—and it was revolutionary in the sense that the way in which the zoos and related conservation organizations decide to wage their war was based not on competition, but on cooperation. Of course, zoos, universities, and big conservation organizations had been cooperating on projects to some extent for years, but never had they sat down together, analyzed the needs of an entire country, prioritized those needs, and then pooled their resources in an effort to solve the most important problems systematically.

Madagascar is the fourth largest island in the world. It lies in the Indian Ocean, some 248 miles away from mainland Africa, and it was formed not by volcanic activity but by a process scientists call "continental drift." Some 165 million years ago, Madagascar pulled away from Africa, and life on this miraculous island developed largely in splendid isolation from the rest of the world. It is one of the most ecologically rich places on earth. It has 25 percent of Africa's flowering plants, and half of the world's chameleons are found there. In fact, almost all of Madagascar's reptile and amphibian species, half of its birds, and all of its lemurs are endemic to the island, meaning that they are found nowhere else on earth.

Lemurs are small prosimians, or, roughly speaking, pre-primates. They don't look much like monkeys, though. The best description might have them as sort of a cross between a monkey, a cat, and a raccoon. There are at least forty different species of living lemurs, although we keep finding more—another three in 2000—and our zoo, in conjunction with a brilliant geneticist from the Henry Doorly Zoo named Edward Lewis, continue to find evidence that there are even more species yet to be identified.

They range in size from the tiny mouse lemurs, which weigh about an ounce or so, up to the fifteen-pound indri lemurs and sifakas, which are about the size of a large house cat. Most lemurs live in trees, and the majority are also diurnal, meaning they are active during the day, but many other species prefer the safety of the night. They are primarily vegetarians, but again there are exceptions and insects can be an important part of the diets of some species. They are beautiful, gentle, intelligent creatures, and people who get to know them almost can't help falling in love with them. Of all of Madagascar's endangered flora and fauna, it is lemurs that have the most charisma.

I fell in love with them when I first arrived at the Indianapolis Zoo. Most of their lemurs were ring-tails, beautiful light-gray animals having long, erect tails with alternating black and gray bands. They are diurnal species and one of the few that live on the ground. At Indianapolis they were displayed on an island that keepers accessed via an underwater bridge. Lemurs, like all primates, have an opposable thumb. This simply means that, like you and me, they can touch the tip of their thumb to the tip of each of the rest of their fingers. Their hands are not quite as dexterous as ours, though. Anyway, I would walk across the underwater bridge with a cup full of raisins and some thin slices of apple and, on my arrival, they would leap up on my shoulders and gently take an individual raisin from my outstretched hand. Sometimes they'd sit on my head and look down into my eyes from above, curious, but never threatening. Of course, when the raisins ran out, they quickly lost interest in me, but I never did lose interest in them.

Like virtually every other wild thing living in Madagascar, lemurs are in trouble. Ten species are critically endangered, seven are endangered, and nineteen are considered vulnerable. They are endangered because the people of Madagascar have cut down over 80 percent of their island's forests. In addition to massive habitat loss, many of the larger species are also hunted for food. It is hard to imagine the devastation that humans have brought to what many ecologists describe as an Eden, but one way is to look at satellite images of the island. A huge brown ring in the waters of the Indian Ocean surrounds it. That brown ring is the soils of Madagascar, washing away forever into the sea, the result of unsustainable agricultural practices like slash and burn. Slash and burn, or *tavi,* as it is called on the island, involves going into the forest and cutting down all of the trees. The trees are left to dry out and then are lit on fire. The ash from the burned wood provides fertilizer for the crops, but the effect does not last long. The soil is quickly depleted and the farmers move on, deeper in the forest, repeating the cycle. All lemurs are now protected by CITES (the international treaty that restricts the trade in endangered species), and it is no longer legal to hunt them or capture them for trade unless it is for scientific purposes or to breed them in zoos. But still their numbers decline precipitously.

Acutely aware of the threat to this unique island, the San Francisco Zoo, the Durrell Wildlife Conservation Trust (headquartered in Jersey, a part of the British Islands), the State University of New York at Stony Brook, Duke University, the Saint Louis Zoo, the Zurich Zoo, and others have banded together to develop a plan for conserving wildlife in Madagascar, focusing on lemurs as the flagship species. At first, we worked closely with the only two zoos in Madagascar, Parc Botanique et Zoologique de Tsimbazaza (or the Tsimbazaza Zoo) and the Parc Zoologique d'Ivoloina (or the Ivoloina Zoo). Over the next fifteen years, however, the emphasis shifted somewhat. Although the Madagascar Fauna Group (or MFG, as it came to be known) took over the Ivoloina Zoo and operates it to this day, the group ceased working with Tsimbazaza within a few years and shifted their focus more to in-country conservation. Those conservation efforts were destined to be far more successful in conjunction with Ivoloina because the zoo was located close to a natural area named Betampona.

Located some thirty miles away, Betampona has been called the most important patch of eastern lowland rain forest left in Madagascar. In 1987 the cooperating organizations agreed to each donate nine thousand dollars per year, hire two wonderful Americans, Andrea Katz and Charlie Welch from Duke University, and rebuild the zoo as a site for captive breeding of endangered species and for local conservation education. The zoo reopened in 1990 and began to work closely with the Betampona nature area. Eventually this powerful partnership was to result in one of the most exciting projects in the history of Madagascar conservation, the release of captive-born black-and-white ruffed lemurs back in with the remnants of the wild population at Betampona.

So what is the MFG today? What are we doing, and why is it so important? Well, today the MFG is a powerful collaborative with a budget in excess of a quarter of a million dollars. It has about forty members, with about two dozen from America (including zoos, universities, and natural history muse-

ums), about half as many from Europe, and members from Africa and
Australia. In cooperation with the CBSG, the MFG has done an exhaustive
survey on lemurs throughout Madagascar and developed priorities for biomedical survey work (done largely by the Saint Louis Zoo) and genetic work
(done largely by the universities and the Henry Doorly Zoo). The group has
also undertaken a Conservation Action Management Plan (or CAMP) in conjunction with CBSG. The management plan looks at virtually every group of
threatened species—from reptiles to amphibians, fish, birds, and mammals—
and develops clear priorities for working with each group in each major
region of the country.

Several of those priority needs are being addressed, either by the MFG directly or by partners of the MFG acting alone or in cooperation with other
partners. For example, a group of French zoos, led by the zoo in Mulhouse,
France, work with the Wildlife Conservation Society to protect endangered
lemurs in the Sahamalaza region. The Wildlife Conservation Society works
extensively with the radiated tortoise and cooperates closely with the Zurich
Zoo in conserving wildlife in the Masoala region, and, perhaps most exciting
of all, the MFG itself is working with another of the original partners, the
Durrell Wildlife Conservation Trust, in preserving the giant Malagasy jumping rat in the Menabe region of Madagascar.

I think this program is particularly exciting because, in many ways, it embodies the true potential of collaboration. The Menabe region, located toward
the middle part of the island, is the only place in Madagascar where you can
find the giant jumping rat, the flat-tailed tortoise, Berthe's mouse lemur, and
the narrow-striped mongoose. It is, in other words, the last refuge of four of
the most interesting animals you can imagine.

Durrell worked closely with the government to establish a reserve in
Menabe that is over sixty thousand acres in size. While that's wonderful, they
have also proposed a very clever plan that, if implemented, would increase
the practical size of the reserve by another twenty thousand acres. This is
what they came up with. There are eight villages surrounding this new reserve. Each of those villages has what we might call "common lands," abut-
ting the reserve. Durrell went to each village and offered them a deal, saying,
in effect, "If you will set aside all of your lands adjacent to the new reserve
and agree not to hunt on the land, not to log the land, and not to farm the
land, we'll pay as much as you could have gotten for the timber alone." Then
they came to the MFG and asked, in effect, "What if we can find eight partners, each willing to put up five thousand dollars a year, pay the villagers in

every village surrounding the reserve not to use the land *and* to monitor land use in order to ensure that the villagers are living up to their end of the bargain?" Everybody wins. First, even the smallest zoo can afford five thousand dollars, and for that amount, the zoo could say that they have their own reserve in Madagascar, *plus,* in combination with seven other cooperating zoos, they would be responsible for increasing the effective size of the Menabe Reserve by one-third. The villagers win because they get just as much money as they would have if they had logged their lands *and* they don't have to destroy their environment to do it. The MFG would win because we would get eight new members, who may never have participated in a conservation project in Madagascar otherwise.

I recently returned from Madagascar, having spent sixteen wonderful days looking at the two major projects, the zoo and the reserve, which the MFG manages largely on its own. I had heard all of the things I just described about the island and our work there over the years, but I had never actually seen it for myself. Forgive the hyperbole, but I was blown away.

It takes about a day and a half of flying (with all of the layovers) to get to Tamatave, where the zoo is located, on the eastern coast of the island. I was met by our curator of primates, Ingrid Porton, and by Dr. Karen Freeman, an adventuresome and friendly field expert with a Ph.D. from Oxford, who now works for the MFG. Our first trip was to the zoo. In retrospect, I guess I didn't really know what to expect, but what I saw amazed me. The zoo covers a little less than one thousand acres, and it is everything a modern zoo should be—and more. Beautifully maintained paths lined with gorgeous tropical plants meander around a small lake. Lemurs live free on the grounds, receiving daily food supplements from the staff. Other lemurs live in spacious enclosures but are not allowed out. The lake has a large peninsula jutting out toward the middle. Endangered waterfowl float idly by the shore, and the peninsula can be reached by a lemur-proof bridge.

It is lemur-proof because it is, in effect, a lemur boot camp—a place where lemurs can be released in preparation for returning to the wild. Visitors, if they're lucky, can wander around the peninsula and, perhaps, catch a glimpse of an animal that was born in a zoo in the United States or Europe and is now "in training" to go back to the wild.

The zoo is really the only place where Malagasy can go, at least if they live in the city of Tamatave, to wander in anything like what we would call a park or a forest. Most of the zoo is really more of a nature preserve, but they demonstrate reforestation techniques and display new plants and modern agricultural cultivation techniques that are far less detrimental than the old techniques of cutting and burning the forests. Trails wind along ridges that offer breathtaking views and along the stream that feeds the lake, a stream that features a lovely waterfall marking the trails' end.

They have a beautiful school, where they help prepare young children for the examinations that allow them to enter both secondary school and college. My visit apparently required a tonga soa, or formal welcome, and the children had organized a program of songs and dances to honor and welcome me. They kept referring to me (in French) as the Grand Chef. (Even though I don't know a thing about cooking, I currently chair the MFG, so I guess that's what they were referring to.)

It is, I suppose, a little piece of paradise for Tamatave, the third largest city in Madagascar. But it also brings into sharp relief the problems that plague the entire island. The road leading up to the zoo had been largely washed away in a recent flood. Electricity still has not quite reached the zoo grounds yet, although the utility poles end not too far away. Andrea and Charlie had spent much of their eighteen years with the MFG trying to convince the local officials to bring power the last few miles to our facility. The most telling thing, though, was the sights that greeted me as I traveled along the bumpy dirt track that led to the zoo. Alongside the road there were dozens of woven sun shades. Underneath each one sat a man, or a woman, or even a young child. Next to each person was a pile of large rocks. They spent their days, all day, every day, with a small iron hammer, breaking the large rocks down to gravel to sell to contractors for construction projects. I guess it would be helpful to put the economic problems of Madagascar in the context of those endlessly swinging hammers.

It would make sense to you if I told you that I make in one day what those people make in a year. It would be almost, but not quite, incomprehensible. Sadly it is not true. I make in one day what they make in *ten* years. That, I think, cannot be comprehended. That, although true, is too much to believe.

And so my days went, a jumble of awe, astonishment, and wonderment as I experienced the ecological richness mixed with the rude reality of a nation that struggles to feed its people. Our next excursion, this time up to the Betampona Reserve, brought those same wild contradictions to the fore. To get to the reserve, you travel another thirty-five kilometers down the washed-out road. The road ends abruptly where a bridge once crossed the Tamatave River. The bridge was washed away in 2001, so now you have to cross the river in dugout canoes. There is another fifteen kilometers of road, but only one

vehicle ever travels that road. It is a small pickup truck owned by an enter-
prising man who took his truck apart at the end of the road and carried the
pieces across the river by canoe to be reassembled on the other side. He has
a monopoly on motorized transport from the site of the old bridge until the
end of the road. If you want to ride, you ride with him and pay his price.

We had arranged to have him take us the last fifteen kilometers to where
the trail up to the reserve began. He did, but he showed up two hours late.
After cramming all of our food and supplies onto his little truck (along with
ourselves and anybody else, it seemed, who might need a ride down the
road), we drove to the end of the road and then began the long, uphill climb
to the reserve. Betampona (which, I'm convinced, translates to something
like "the place where all paths go uphill, even if you're returning from whence
you came") is Madagascar's oldest reserve, founded by the French almost one
hundred years ago. Like many of Madagascar's most important reserves, it is
stretched out along the tops of the mountains, because those are the most in-
accessible places to try and cut the forest.

The climb begins in a squalid little village that is marked by a few small
stores that sell basic items and a shop or two with the Malagasy equivalent of
fast food (fried potatoes and roasted corn). The path leads through and
around a number of other small villages, located alongside fields of rice, cross-
ing about fifteen rivers that race down from the mountains above. After a few
hours of walking, you reach the real climb, an exhausting, nonstop, straight-
up trek. We had porters, or I'd still be struggling up that mountain.

At the top is another one of those odd surprises. At the entrance to the reserve is the village of Rendrirendry. Everyone who lives there has someone
employed by the MFG in their household. Of course, they make a very good
living working for us (at least by comparative standards—we pay about
eighty dollars per month). But this village is different for so many other reasons. The ground is covered by a carpet of grass and bougainvilleas, and
amaryllis bloom everywhere. The houses are sturdy, and the offices of the MFG
are built of cement. There are signs of technology everywhere—radios, GPS
units, and camera traps seem to cover every table.

The real surprise, though, is that everyone in the village is happy. The children greet us with another tonga soa, and I take my turn pounding a hollow
tube of bamboo on a low bench of wood in time with their singing. Mothers
play with their youngsters while fathers weave new mats or relax in the shade
of huge trees. After everything you see walking up here, it looks like the paradise you had always hoped to find.

And that is only the beginning. The trail from the village leads into the reserve, and within minutes we see lemurs leaping gracefully high in the trees. A pair of mongoose walking along the trail greet us with surprise, but not alarm. They are brown mongoose, and my book on the mammals of Madagascar does not contain a photograph of them. Dr. Freeman tells me that there simply are no good pictures of them because they are so rarely seen. The ancient forest seems so totally alive—it embraces us with a hot warmth that is almost sensual. (It also sheds a few forest leeches down my shirt. When they are fully engorged with my blood, they drop off, but I continue to bleed because they excrete an anticoagulant. By the end of my first hike, I will look like I came out on the short end of a bad encounter with a machine gun.)

Finally our forest agents spot the lemurs wearing the radio collars. They are the ones we have come to see. They are the ones that were born thousands of miles away in a zoo and now live where they belong. If the forest is a cathedral, then surely these are the answer to a prayer.

So what am I most proud of? Well, several things. First, at the Ivoloina Zoo, I am thrilled by the variety of educational programs we offer. For middle school students, we have courses designed to help them pass their national exams, so that they can get into high school. We also offer classes in environmental education and have camps during vacation periods. Similarly, we help prepare high school students for the national exams required to get into college. Teacher training is a big part of our ongoing efforts, with a focus on elementary school teachers. But we also offer courses for adults. For example, we've created an ecologically friendly agroforestry nursery with native trees as well as exotic species that can be planted as a cash crop. We also show area farmers how they can improve techniques for rice cultivation, plant alternative fruits and vegetables for the market, and use specific plants to help control erosion.

In the zoo, we keep critically endangered species of lemurs in breeding groups, plus we serve as a repository for confiscated animals liked tortoises. This means that we have viable populations for reintroductions or relocations. We can also use the zoo to help train professionals in animal management, veterinary science, and field biology.

In the reserve, we track the progress of our ruffed lemur restocking program, and this effort has expanded into a training program for conservation agents and, more broadly, into lowland rain forest research in general. We now have Malagasy nationals studying phenology (the cycle of plant blooming and fruit production), as well as Malagasy professionals who specialize in monitoring reptiles, amphibians, birds, and mammals (including, of course, lemurs). We're also training young college students, all of whom are learning the skills of field biology as they advance through their studies.

There is much more to be done to achieve the MFG's vision of mobilizing zoos, aquariums, universities, and conservation organizations worldwide in an effort to halt the extinction of the unique wildlife of Madagascar, but we are making enormous progress. The MFG works because we can give even the smallest zoos and aquariums "sole ownership" of a specific project, while allowing them to share the credit of what we all do together. The projects have excellent PR value and can be presented to the visiting public of each collaborating institution in a way that shows visitors that their support can, and does, make a difference. Even the smallest zoos can be a part of these efforts. Because we are required to demonstrate to the federal government that we are making a difference in field conservation every time we bring even a single endangered or threatened animal into the country, this type of collaboration can make the permitting process far easier than it is with, say, giant pandas. From a simple business point of view, through cooperation, I can have field personnel working in Madagascar for a fraction of the cost. Since we have full-time staff already in place, we can help zoos with their own specific projects and, more importantly, steer them toward projects that we all agree have the highest priority. Finally, because the MFG is registered with the government, we already have what you might call mini-treaties in place with different ministries of the government of Madagascar—the Ministries of Environment, Education, and Research. This makes working in Madagascar far, far easier than otherwise.

All of this contrasts vividly with what happened with the giant panda in China. I guess the moral is that while civil wars can be horrible, a little revolution now and then can be a very good thing.

15

This Would Be a Nice Place—
If It Weren't for the Visitors

On August 30, 2003, the *St. Louis Post-Dispatch* printed a letter to the editor that's worth reprinting here.

> I noticed in Deb Peterson's July 31 column that the Saint Louis Zoo is studying whether to expand the area of taxation to support the Zoo's programs. While I love our Zoo and am proud of its programs, I would not personally support it until I can once again visit it in safety. I have contacted Zoo administrators at least three times, requesting that they discontinue the sale of balloons, which put latex into the air and are dangerous for anyone with a latex allergy. Balloons are lovely and fun, but they are not a necessary part of Zoo operations. I discontinued my Zoo membership, since I can no longer visit, at least not on busy summer days. I would definitely object to being taxed to support an institution so unresponsive to a legitimate medical request.

I was, I have to admit, one of the people the author of the letter talked to about her latex allergy. I didn't point out that many people (including myself) are allergic to cats. If I had pointed that out, I probably would have gone on to say that my allergy isn't a good reason for us to stop our work with Siberian tigers. I did say that the balloons actually generated quite a bit of revenue for the zoo and that the vast majority of our visitors seem to enjoy them. I don't remember what else I said, but I'm sure I was unfailingly polite. I'm also sure that the whole time I was talking, I was thinking to myself, "This would be a pretty nice place—if it weren't for the visitors."

What makes a zoo a zoo, as opposed to a theme park on the one hand, or a research center on the other hand, is our mix of missions. We are, most certainly, serious scientific institutions. But no one visits research institutions, or at least very few of us do. People do visit theme parks, but with the exception of Disney's Animal Kingdom, the Sea World parks, and a handful of others, they don't afford visitors an opportunity to learn much, if anything, about the world around them. The fact that zoos fall somewhere in between is our greatest strength and, at the same time, our greatest weakness.

In an article for the *Washington Post,* Jeffrey Hyson, who is writing a cultural history of zoos, talks about the early formation of the National Zoo. At the time of its founding in 1889, the advocates for the zoo envisioned an essentially "scientific" enterprise. It was to be largely closed to the public with only a few acres open to visitors—in other words, a research center that hosted a few guests. That ideal was never really given any practical consideration. The people of Washington (and, more importantly, the congressmen who were asked to provide the funding for the new zoo) didn't want to see what Hyson calls "charismatically challenged species" such as pronghorn and elk, both of which were in dire need of conservation at the time. They wanted to see exotic species—lions and tigers, elephants and apes. As it turns out, the National Zoo does have a wonderful research facility that is, indeed, closed to the public. It's called Front Royal, and an incredible amount of conservation research and breeding are accomplished there. But the zoo itself is more or less entirely open to the public. From a research and conservation point of view, having all those people coming in makes life very difficult. But we're not just about that. We're also about educating people or, to my way of thinking, getting people to care more for living things. Obviously, this means we want people to visit; in fact, having them visit is a very large part of the reason for our continued existence. Still, there are days that I have to remind myself that if it weren't for the visitors, I'd probably be out of a job.

People visit zoos for a variety of reasons. Certainly one of them is an opportunity to see animals that they would otherwise never get to see. Like any museum, there is a premium placed on seeing the real thing. There are other reasons, though. It is very much a social experience. Few people actually come to our zoo alone. They come with friends and family. And they spend an enormous amount of time interacting with one another—more time than they actually spend looking at animals. When asked, many visitors will tell you that a zoo visit is an educational experience, although not nearly enough of them pause to read the information that we spend so much time carefully preparing. It is also a rich aesthetic experience. Plants and animals are beautiful to some of us in the same way that a painting in an art museum is beautiful. Finally, for many visitors it is a reverential experience. For some of us, being in the presence of magnificent animals can evoke the same kind of awe and wonder you get when stepping into a magnificent cathedral.

The reasons people come to zoos are very important, for people come in huge numbers. By the end of the last century, 1999, there were 176 accredited zoos in America. That year, nearly 127 million people visited a zoo in our country. That's more people than attended all professional baseball, basketball, football, and hockey games *combined*. Today, there are 214 accredited zoos, and national attendance is up to 142 million. Designing zoos so that their exhibits and programs meet the needs of those visitors, while still meeting the needs of the animals and serving the research and conservation goals, is arguably our greatest challenge.

As an example, let's take an animal that is found in most major zoos, the African lion. Most people have a fairly good idea of what life is like in the wild for lions. They live in groups of varying sizes called prides. The received wisdom is that there is one male in the pride and that younger male cubs, as they approach maturity, are driven out of the pride by the dominant male. In fact, we've seen many instances of two related males taking over a pride in recent years.

Males are good hunters, but their primary job is to defend their group of females from other males who have grown large and strong enough after much time on their own to challenge a male for control of the pride. The popular idea of the lazy male, who lets the females do the hard work of hunting and who eats his fill of the kill before the females and cubs can approach, is largely true, but it's only part of the story. Defending the pride and the pride's territory is serious, dangerous work. Male lions in zoos are handsome, regal animals. Male lions in the wild are grizzled, and their faces are often covered with terrible scars. Defending the pride is often a matter of life or death for the males.

When a male does successfully take over a pride, the pregnant females will spontaneously abort and the male will usually kill any young cubs that were sired by his predecessor. The females will then go into estrus almost immediately and the new dominant male will breed them, assuring that all the babies will be, genetically speaking, his. By the way, lions must hold some sort of record for mating. They copulate every ten minutes or so when the female is at the height of her cycle and have been known to copulate over 150 times in a day.

The size of their range varies greatly according to the food supply. I once watched a pride that had killed a hippo feed for three days straight, and they never once moved over twenty yards from the site of the kill. In other words, if given a steady supply of food, lions don't need much space at all. On the other hand, if food is scarce, lions might cover several miles in the course of one night's hunting.

How does all this contrast with life in the zoo? For one thing, their range is severely constricted. Even where food is abundant and competition from other prides is not present (a situation that would never last for long in nature), no pride in the wild would occupy as small a space as is allocated in the average zoo. Lions in zoos do continue to hunt—for example, ours regularly go after our free-ranging guinea fowl—but their food supply in zoos is assured. Ours are fed a mixture of ground horse that includes the bones and organs along with vitamin supplements. They fast one day a week, in contrast with life in the wild, where they may have to go several days without eating, depending on the availability of game and their success in bringing it down.

Their immediate environment is much richer, though. Keepers add giant balls, rubber swings on chains, logs, giant chew toys, or whole bones from cattle to their exhibit on a rotating basis. Lions are scent-oriented, so keepers might spray different perfumes, hide spots of aromatic kitchen spices, or bring in dung from prey species and hide it around the exhibit yard. In many zoos—ours is an example—animals can come inside at night or stay outside depending on their own preference.

Our lions mark their territory regularly by spraying (sometimes hitting a visitor during the process) just as they would in the wild. They tend to largely ignore visitors, but they may attend to the presence of a keeper briefly. When mating, they mate as often as they would in the wild. Vet care for lions is extensive. They're given regular exams, with full blood workups, receive inoculations that guard against diseases that would strike them in the wild or in human care, and get dental check-ups more often than many people do. In fact, I once watched our vet staff do a root canal on a lion's canine, a pretty amazing procedure. As a result of their medical care and diet, plus the fact that they don't have to risk life and limb bringing down a running zebra,

One of the Saint Louis Zoo's rescued sun bears.
JOHN STORJOHANN

handle, he was turned over to the government. The female was also confiscated by the government, along with seven other sun bears, all of which were then turned over to the Sepilok Rehabilitation Center, an orangutan forest reserve located in a place called Sabah, Borneo. The nine sun bears were all held in small enclosures, and the director of the reserve, Dr. Bosi, was deeply concerned for their welfare. The reserve was never intended to house sun bears, and it had neither the experience nor the resources to deal with them.

Dr. Bosi called the American Zoo and Aquarium Association and asked for our help in finding them good homes. It wasn't the first time that he had been faced with essentially the same problem. The sun bear Species Survival Plan, which was managed by Judy Ball at the Woodland Park Zoo in Seattle, had brought ten sun bears over to America in 1996. By May 2000 our two bears, along with eight others, were set to make a similar journey. Cheryl Frederick, who had by then replaced Judy Ball at the Woodland Park Zoo, handled the transfer of the ten bears. Two came to St. Louis; Woodland Park, San Diego, Houston, Cleveland, the Minnesota Zoo, the Lincoln Park Zoo in Chicago, and the Gladys Porter Zoo in Brownsville, Texas, agreed to adopt the rest of them.

Our sun bear exhibit was nice when I arrived at the zoo, but Steve Bircher, the curator of carnivores who had rescued our grizzly bears, Bert and Ernie, earlier in his career, had already made several improvements. The new exhibit had a stream for the bears to play in, climbing trees, and changing dead logs for the bears to paw through in search of small insects. The holding area for the bears had not been changed at all, though, and at the time our keeper was mauled, it was not much different from any other overnight holding area. There were separate dens for each animal, which were connected to the outdoor yard by a passage equipped with a gate. To let the animals out in the morning, the gate was opened, the bears ambled out, the gate was closed behind them, and then (and here's the critical point) the keeper walked out to the front of the exhibit and counted the bears. If there were two bears outside, that meant that there were no bears inside. If there were no bears inside, then it was, obviously, safe to go into their dens and clean. It all sounds pretty simple, and in fact it is.

The keeper who was mauled has worked for the zoo for eighteen years. He doesn't make mistakes often. In fact, he hadn't made a single significant mistake in all those years until that morning in February, when he made not one, but two errors in a row. What caused the errors was that, on the morning in question, he wasn't working alone. He was training a new keeper. So instead of walking around the front of the exhibit to check and see if both bears had moved through the passage, he simply called to the trainee and asked him if he saw the bears go out. The trainee replied that the bears had moved down the passage, but neither man actually walked around to the front to ensure that they had moved all the way out to the exhibit. Our female had, but our young male hadn't. He was still in the passageway that connected to his den. So that was mistake number one.

Mistake number two was that neither keeper looked closely into the den areas before walking in to clean. That turned out to be okay for our trainee, as the female was safely locked out in the yard. For our keeper, though, it was another matter entirely.

Now this particular keeper is big. I mean really big. He stands a good six feet two inches and probably goes about 240 pounds. That would make him about twice the size of our young male bear. But what our keeper had on the bear in size, he lacked in teeth and claws. As soon as he stepped in the den, he saw the bear and knew he had a problem.

A sun bear is just as likely to attack an intruder in his home as you would be in yours. That's not to say that he will, necessarily. Just like you might flee from your home if you discovered an intruder, our bear could have fled into the locked passageway. I don't know exactly which I would do if an intruder suddenly walked into my home and, truth be told, you probably don't, either. Scientists refer to this particular dilemma as "flight or fight," and it's simply impossible, under realistic circumstances, to predict which way a bear (or a human) will go. Our keeper's bear picked "fight" and went for his ankles.

It happened so fast that he was knocked off balance and fell to one knee. The bear, still low to the ground, bit through his knee. The keeper regained his footing, and the bear got one last bite to the keeper's forearm as he pushed the bear away; then, staggering backwards from the force of the attack, the keeper closed the door to the den, leaving the bear safely inside. It all happened so fast that our trainee was able to do nothing but grab his water hose and turn it on the bear. All that managed to do was get our keeper wet. Given the man's size, I'm afraid he pretty much filled the entire doorway, so the trainee had very little room to squirt.

I wouldn't be so flippant about all of this if our keeper had been seriously injured. He wasn't. Despite the bear's long claws and sharp teeth, he didn't nick a vein or artery. Our keeper spent the better part of the day in the emergency room and was stiff and sore in the days to come, but he recovered completely.

When I first heard about what happened while talking on a satellite phone thousands of miles away, I was upset. Bill Boever, who was in charge while I was away, had broken it to me diplomatically: "Don't worry, he's all right, but a senior keeper got attacked by a sun bear." He also quickly pointed out that the bear didn't get out, and no one else suffered any harm. After that, the story tumbled out pretty much the way I've described it. Of course, the media came out in full force. In fact, the TV crews were on the scene before our keeper even left for the hospital to have his wounds inspected and cleaned. No big surprise there—they listen to the emergency calls, and we happen to have one TV station that's virtually within walking distance of the zoo. We expected the media coverage, but what we didn't expect was that PETA would send representatives to St. Louis to call a press conference.

PETA is the acronym for People for the Ethical Treatment of Animals. It is a relatively small not-for-profit (its annual budget in 2002 was about $17 million—small in comparison to our $40 million annual budget) that is dedicated to animal rights (not welfare—a distinction that we'll talk about in just a second). PETA operates under the simple principle that animals are not ours to eat, wear, experiment on, or use for entertainment. They focus their attention on farms, especially "factory farms," the use of animals for medical testing, the fur trade, and the use of animals in the entertainment industry. It is probably fair to say that the majority of PETA's efforts go to the generation of publicity. In fact, Ingrid Newkirk, one of PETA's cofounders, was quoted in *USA Today* as saying, "probably everything we do is a publicity stunt." (A better quote comes from the *New Yorker*—"We are complete press sluts.") Most of their money comes from donations, but I doubt that most people are aware of where their money really goes. For example, PETA does run animal shelters and spay-neuter programs for pets. This is an unenviable task. Last year alone, PETA euthanized about as many animals as the Saint Louis Zoo has in its entire hundred-year history.

So after our keeper had been injured, what happened next is pure PETA. First, they said that our bear suddenly snapped because of his years in captivity. That's the kind of lie that's difficult to counter because it is made up of two separate truths. Yes, it's true that the bear, whose name is Rimba, has lived here for years, and yes, it's true that he suddenly attacked our keeper. But the two things aren't in any way connected. I can guarantee you that, if you were dumb enough to try it, you could walk into any den, in the zoo or in the wild, occupied by any bear, anywhere in the world, and one of two things would happen. Either the bear would find a way to run away, or the bear would attack you. But the spokesperson for PETA would lead you to believe otherwise. That just angers me. Their second lie angered me even more. They said "zoos often fabricate rescues to make wild captures more palatable to the public." It should be obvious, from my description of what usually happens to these animals, that the handful of the remaining eight hundred sun bears that make it safely to a refuge are not getting there because zoos are "fabricating rescues."

But while the lies merely anger me, what infuriates me is when they go on to say that their organization would prefer that bears go extinct rather than have them live in captivity. Perhaps *infuriates* is the wrong word. Maybe *sick-*

ens is closer to what I feel. I should save the word *fury* for what I feel when I realize that their organization does absolutely nothing to help save animals in the wild.

Leaving aside for the moment the question of whether People for the Ethical Treatment of Animals behaved, in the case of the our sun bear, in an "ethical" fashion themselves, let's take a broader look at what ethicists have to say about the relationship between people and animals. The first critical distinction we have to make is the difference between animal *rights* and animal *welfare*.

Zoo professionals have a very technical definition of animal welfare. I'll give you the technical definition, but it might be better to say it in plain English first. Basically what the definition says is that animal welfare is a combination of an animal's physical well-being and an animal's psychological well-being. If you wanted to apply that definition to your family dog, you'd probably begin by first ensuring that Fido had a good diet, plenty of water, a clean bed, good medical care, and so on. Then you'd ask, simply put, if his tail was wagging. The technical definition of animal welfare is "the degree to which an animal can cope with challenges in its environment as determined by a combination of measures of health (including preclinical physiological responses) and measures of psychological well-being." Physical health is relatively easy to measure. Basically, what we do is ensure that the animal doesn't have any diseases or any other bad physical conditions that result from things like inadequate nutrition, lack of exercise, or the lack of appropriate medical care.

Psychological well-being is much harder to measure. The truth is, even if Fido's tail is wagging, it doesn't necessarily mean that he's happy. In fact, we don't really understand what being truly "happy" means, even for ourselves, much less for dogs. For example, I hate writing. If I had a choice between sitting here in my office on a gorgeous Sunday, pounding away on this computer keyboard or playing golf, my guess is that playing golf would be the behavior I'd pick. But here I am, writing instead of playing golf (even though it's gorgeous outside and there's a golf course about ten minutes from here). If you really plumbed the depths of my psychological well-being, maybe you'd find that going through the aggravation of writing all of this will eventually make me happier *someday* than playing golf would *today*. Maybe I'm relatively happy even though I'm doing something that I really don't enjoy. More likely, though, is that I simply don't know what makes me happy, at least not on this beautiful Sunday.

For zoo professionals, psychological well-being is dependent on the opportunity for animals to perform strongly motivated, species-appropriate behaviors. In my case, I'm writing. I'm strongly motivated to do so, I have an opportunity to do so (my wife said I didn't have to work in the garden with her today), and even though there might be other things I want to do, I'm writing instead. In other words, if you asked me if I was happy to be here, I'd say, "no." But if you asked a zoo professional if I was happy, she'd say "yes." I have an opportunity to perform a range of strongly motivated behaviors, and I picked writing over golf. Maybe I'm happy to be doing this and just don't know it.

How does this definition play out with Fido? Well, that depends. Let's say you have a golden retriever. Let's say your retriever is strongly motivated to fetch things and bring them back to you. Further, let's say you've thrown your back out and can't toss him a tennis ball. His tail may be wagging when he sees you, but he doesn't have an opportunity to engage in what he'd really like to be doing—in this case, fetching a ball. The only way we have of ensuring psychological well-being is making sure that animals have choices that they can make in response to varying physiological states (for example, if a bull elephant is in *musth,* does he have an opportunity to breed), developmental stages (for example, if a baby bird is approaching the time to be weaned from its parents, does it have an opportunity to learn to fly), variable environmental conditions (for example, if a snake needs to regulate its body temperature, can it choose between warmer or cooler areas within its enclosure), and social situations (for example, does a chimpanzee has an opportunity to interact with other chimpanzees and create or move within a social hierarchy).

Zoos are animal *welfare* organizations. They are committed to ensuring that animals in zoos are physically healthy (or at least that no physical harm comes to them because of something we did or something that we shouldn't have done) and psychologically healthy (at least to the extent that we can provide them with opportunities and choices that allow them to do what they're strongly motivated to do and allow them to have normal choices with regard to changes in their physiology, life cycle, external environment, and social conditions). Zoos are also committed to essentially the same things for animals in the wild. In particular, we don't want to see their physical or psychological health compromised by humans—something that happens, for example, when we destroy their habitat, introduce new and noxious plants or animals to their environment, poach them, pollute their habitat, and so on.

All of this is different from the idea of animal *rights*. Animal rights is a philosophical belief system based on the idea that animals should have the same moral rights as humans and the same inherent value. This notion suggests that, just like you, animals should have the right not to be killed and eaten as food. Like you, they should have the right not to be used for experimentation or entertainment. Like you, they should have the right to be free.

Animal rights, as a philosophy, is relatively new. In 1975 an Australian philosopher named Peter Singer published a book called *Animal Liberation,* which challenged the notion of human dominance over other animals and articulated the idea that other animals should have the same rights as humans. Singer said, "Human beings have come to realize that they [are] animals themselves. It can no longer be maintained by anyone but a religious fanatic that man is the special darling of the universe, or that other animals were created to provide us with food, or that we have divine authority over them, and divine permission to kill them." Robert Wright, writing for the *New Republic* in 1990, summarized it this way: The core of the animal rights philosophy is the dismissal of differences between people and animals—language, reason, morality, free will—as ethically irrelevant. Animal rights asserts equal moral status to all living things based on the ability to feel pain. In this ethic, all human use of animals—for food, clothing, sport, companionship, or medical research—is "speciesism," the moral equivalent of racism.

On the face of it, this philosophy doesn't appear all that unreasonable to many people, but there are many problems if you follow it out to its logical conclusions. First, it should be a lot easier to determine if living things feel pain than it is to determine if they're happy. I'm not too sure if I'm happy or not right now, but I'm quite sure that, if I pounded my finger with a hammer, I would feel pain. I think chickens feel pain when they're slaughtered for food, as do cows, pigs, or any of the other animals we eat. That would include fish and frogs, too. I suspect it would include lobsters and snails. In fact, it may well include chocolate-covered ants. So far, so good. If you adhere to this philosophy, you should become a vegan, a person who eats no animal food at all. Not surprisingly, PETA, which is after all a strict animal rights organization, opposes eating meat or animal by-products like milk.

I personally think you'd be okay with plants, but I'm not really sure about *all* plants. I remember, on my first trip to Africa, our guide pointed out that the giraffes we were watching, which were feeding on acacia, were all positioned downwind from the trees they were eating. He then told us why. It would seem that the species of acacia they were eating emits a distress chemical when being munched. That chemical causes other acacias to produce a chemical that makes the leaves taste awful. The chemical is carried from tree to tree by the wind. As long as the giraffes stay downwind of the trees, the leaves remain sweet. It is silly to think that the acacias feel pain, but it's magical to think that they communicate to one another!

So let's assume that no plants feel pain and that all animals (that we'd be interested in eating) do feel pain. If you can't eat them, you can't really condone using them for clothing, either. Obviously, fur and leather would be out, and according to PETA, silk would be off-limits, too, on the assumption that "silkworms can feel pain."

If animals have the same right not to be eaten as you do, then they also have the same rights not to be used as subjects in medical research. For example, Ingrid Newkirk, the cofounder of PETA and still the group's most important spokesperson, was quoted in *Vogue* as saying, "Even if animal research resulted in a cure for AIDS, we'd be against it." This is going quite a bit further than attacking a cosmetics company for testing their products on lab rabbits. It means that, if you adhere absolutely to the tenets of the animal rights philosophy, you would ban all medical testing using any other animals than humans. I think it would also mean that you would not use any drug or therapy that was devised or perfected using research on nonhumans.

This philosophy also means that we would no longer have pets or companion animals. Ms. Newkirk regards pet ownership as "an absolutely abysmal situation brought on by human intervention." She goes on to say, "One day we would like an end to pet shops and the breeding of animals." Animals like dogs, in this worldview, should be freed to "pursue their natural lives in the wild." This makes sense in the animal rights scheme of things. Keeping a dog or a cat is no different than keeping a slave.

Giving the same rights to nonhuman animals as we do to humans means that we must regard them all the same. That is what Ms. Newkirk means when she says, "A rat is a pig is a dog is a boy." That's why she can say something like, "Six million Jews died in concentration camps, but six billion broiler chickens will die this year in slaughterhouses." It may strike some of you reading this as repugnant when someone places the suffering of the Jews under Hitler as being no different than the suffering of chickens in a slaughterhouse but, if you believe in the philosophy of animal rights, there's no real difference. Keeping slaves is no worse than keeping a pet dog, and killing

Jews is no different than killing chickens. That's why you get people like Michael Fox (at the time the vice president of the Humane Society of the United States) saying, "The life of an ant and the life of my child should be granted equal consideration."

The vast majority of us would never adopt this extreme philosophical stance. In fact, we couldn't even if we wanted to. Wearing plastic shoes, as I'm told Ms. Newkirk does, still doesn't solve the problem of animal suffering. I say this because it takes hydrocarbons (in this case, oil) to make plastic. Oil comes from the environment. Destroying wild habitats in an effort to produce oil causes, at least indirectly, harm and suffering to animals. In fact, the same argument can be made for farming. Every acre of rain forest that is converted to farmland condemns not just individual animals to suffering and death, but may cause whole species to suffer extinction. And, just like it takes oil to make Ms. Newkirk's plastic shoes, it takes oil to make a potato.

In the United States, it takes approximately 22 gallons of gasoline, 203 pounds of fertilizer, and two pounds of chemical insecticides and pesticides to grow one acre of crops a year. Before 1900, using our old methods of farming, we actually got more calories of food energy from farming than it took to grow the crops. Now it takes at least eight calories in the form of fossil fuels to produce one calorie of food. Those fossil fuels have to come from somewhere, and no matter how careful we are at extracting them from the ground, some animals will undoubtedly suffer as a consequence. Even if a potato doesn't feel pain when you harvest it and eat it, some living thing still probably suffered as an indirect consequence.

In any case, from an animal rights point of view, zoos are not morally defensible. They would say that keeping an animal in confinement is no different than keeping an innocent person in jail. Drawing blood from an elephant in a zoo, something that we did weekly and, at times, daily, in order to unlock the mysteries of elephant hormones, causes the elephants as much pain as drawing blood from a diabetic. Even if we are drawing blood to help treat an animal for a health problem like salmonella, we are doing something that, in the animal rights view, is morally wrong.

But there are other ethical perspectives on humans and animals aside from the one we're calling animal rights. Tom Regan, in his essay "Are Zoos Morally Defensible?" outlines several. The philosophy that comes closest to the one I believe in is sometimes called *holism*. Holism as a philosophy can be traced to one of the most important environmentalists of the twentieth century, Aldo Leopold. I'll let Regan describe it for you:

Leopold rejects the individualism so dear to the hearts of those who build their moral thinking on the welfare or rights of the individual. What has ultimate value is not the individual but the collective, not the part but the whole, meaning the entire biosphere and its constituent ecosystems. Acts are right, Leopold claims, if they tend to promote the integrity, beauty, diversity, and harmony of the biotic community; they are wrong if they tend contrariwise. As for individuals, be they humans or other animals, they are merely "members of the biotic team," having neither more nor less value in themselves than any other member—having, that is, no value in themselves. What value individuals have, so far that this is meaningful at all, is instrumental only: They are good to the extent that they promote the welfare of the biotic community.

If you believe in this philosophical approach, it is still not easy to tell right from wrong. How would you consider, for example, my life? I certainly have a detrimental effect on the planet. I drive an SUV, use my air conditioner more than I should, eat fish that come from oceans that are being destroyed by commercial fisheries, and so on. On the other hand, I, and the people I work with every day, do an incredible amount of vital work in saving endangered species around the planet. On balance, are we doing good or are we doing harm?

Tom Regan presents an even better example—the trapping of fur-bearing animals for commercial gain. For someone who believes in the rights of each individual animal, this is clearly morally indefensible. For someone who takes a holistic approach, there's nothing wrong with trapping as long as it doesn't disrupt the integrity, diversity, and sustainability of the ecosystem. Trappers can disrupt the ecosystem, just as the killing of wolves in the United States has caused the deer population to skyrocket in certain areas. On the other hand, if we don't overtrap or overhunt, we may kill individual animals—just not at the risk of damaging the very nature of the environment that supports us all.

According to Regan, when it comes to zoos, there is nothing wrong with keeping animals in confinement, if doing so is good for the larger life of the community. A good example of this occurred in January 2004, when eighteen mountain bongo born in thirteen different US zoos, including the Saint Louis Zoo, returned to their ancestral home at the base of Mount Kenya.

The bongo is the rarest and one of the largest of all the forest antelopes. It's a shy but strikingly vivid animal, with outsized hindquarters and stripes across its legs and body that look like they were painted by hand. It towers over Africa's other antelope, reaching a weight of over nine hundred pounds. Unlike other antelope that react to hunters by fleeing at top speed, bongo often freeze in place, relying on their distinctive striping for camouflage. They can fight back, using their huge horns for defense, but this doesn't help much if they are attacked with guns or bulldozers.

From 1966 through 1975 several bongos were removed from Kenya and brought to American zoos. The remaining animals gradually came to be confined in three tiny habitats—the Aberdares Conservation Area, where we estimate that there are about one hundred animals, the Mau Forest, where we think they are probably extinct (but we're not sure), and Mt. Kenya National Park, where the last wild bongo was seen a little over ten years ago. In other words, from 1975 to 2004, a span of about thirty years, the remaining bongo in Kenya declined to about one hundred animals. Natural predators like the hyena and lion were partly responsible, but as usual, the main culprit was man. Poachers chased them down for meat, and their habitat continued to disappear, forcing the wild bongo to the brink of extinction.

The captive populations in US zoos, on the other hand, thrived during the same period. Although all descended from the original native population in Kenya, coordinated breeding has produced herds in US zoos with high levels of genetic diversity, and there are enough animals to begin the process of returning them to the wild. So in late 2003, select zoos sent eighteen robust animals, including four males and fourteen females, to White Oak Conservation Center in Yulee, Florida, for quarantine. After three months, they were shipped to Mount Kenya Game Ranch, where a special enclosure had been built. The idea was to let the zoo-born animals back into the natural environment under controlled circumstances and then let their offspring out into the national park.

The zoo scientists were thrilled with the result. The bongo left their shipping crates, moved immediately into the thick vegetation inside their huge enclosure, and began to feed as they would in the wild. William Woodley, the chief warden of Mount Kenya National Park, watched along with the zoo vets who escorted the animals over. He didn't say much for the longest time, and then he murmured, "You can take the bongo out of the bush, but you can't take the bush out of the bongo." Mr. Woodley has a special interest in the bongo, by the way. The son of a legendary Kenyan game warden, he is actually nicknamed Bongo Woodley. His father had a special love for the animals, and after the younger Woodley was born, he would use the park radio to check on his infant son while he was out on patrol. Since the radio was for official use only, he had to ask his wife how the baby was doing by using a private code. Every day he would ask, "How's the baby bongo?" The name stuck, and the now-grown William Woodley is known simply as Bongo.

Bongo Woodley has his work cut out for him. Illegal logging continues in the park, and poaching is still a threat, although nowhere near as bad as it has been in the past. Still, we are optimistic that the animals will reestablish themselves, perhaps in as little as three generations.

The question remains, then, "Are we morally justified in keeping bongo in zoos so that their descendants can be used to repopulate the wild?" I believe so. I believe that we are morally justified in keeping animals in zoos even if we think that there is little chance that their descendants will *ever* return to the wild. I see little future for the Malayan sun bear in the wild. Yet I see nothing wrong with keeping a viable population in the United States in the hope that, someday, there will be a better future for them in their native range. Even if there never is, there is the possibility that more people will become aware of their plight and in this, there is hope for their future.

If you believe in the philosophy of holism, you believe it's okay to keep a small number of animals in captivity, especially if it means that, someday, there is a possibility that you can restore the natural habitat back to what it was like before the depredations of humanity. For the record, Mr. Regan disagrees.

His logic is simple. Remember when I said that, in many places, there are too many deer? Mr. Regan says that if there are too many deer, and we recommend that they be culled through hunting, it's fine. But if there are too many humans, we should also advocate that they be culled, too. From an ethical point of view, he says, holism is ethically bankrupt if we don't. In other words, if we're not willing to walk the walk, we can't talk the talk. Most people would consider this an "ivory tower" bit of logic. In fact, we have seen major efforts to limit population size in countries like India and China. Population control in those two places, while not as draconian as lining people up and shooting them, is still undertaken in what can only be described as a most forceful fashion.

But Mr. Regan says that this doesn't count, because human population control is less lethal. Let me quote him exactly. "[W]hy should holists stop short of recommending that the human population be culled using measures no less lethal than those used in the case of controlling the population of deer? Granted, the latter is legal, the former is not. But legality is not a reliable guide to morality, and the question before us is a question of morals, not a question of law. And it is the moral question that must be pressed." Of course, we do cull human populations in a most lethal and, by the way, legal fashion: we call it warfare. Perhaps ethicists like Mr. Regan can help us to develop a more benign form of holism, one that won't require me to be killed for owning an SUV, but will recognize that what I do to sustain and restore the web of life on our planet, in some small measure, mitigates the problems that my very presence creates. That more benign philosophy must encourage others to do likewise.

I do not believe that holism is morally bankrupt, but I insist that the animal rights philosophy is morally untenable. I know that it is for me. I simply cannot agree with PETA when they say that it is better to let a species like the bongo or the Malayan sun bear go extinct rather than to keep them alive in zoos. At the same time, we cannot keep them in zoos if their living conditions are inhumane. That is where the notion of zoos as animal *welfare,* as opposed to animal *rights,* organizations comes to the fore. Zoos are only morally justifiable if they truly provide for the welfare of the individuals in their care, all the while working for the welfare of the species they protect.

This speaks directly to the idea of a more benign holism. In this view, it is essential to act in the best interests of individuals *and* in the best interests of the species. This contrasts vividly with the animal rights perspective, which says that the rights of an individual animal within a species are so paramount that they outweigh the rights of all of the individuals within that species. You may have to read that sentence once or twice again for it to make any sense. In fact, some of us can read it all day, and it will never make any sense. The animal rights perspective leads us down a path of ethical mumbo jumbo that has to confuse even Ingrid Newkirk, a person whose ethics seem, on the face of it, confusing enough already.

I don't say this just because PETA lied about our sun bears, although that was a silly thing to do. I say this because, in 2003, at PETA's request, the AZA and PETA cooperated to rescue seven polar bears from a circus in Puerto Rico. We worked together, and the rescued bears are now thriving in AZA-accredited zoos. Syd Butler, recently president of the AZA, said he would ask the zoo community to do the same again, despite the fact that PETA would seem to be talking out of both sides of its mouth. Why? "Because it's the animals, not public relations games, we care about."

I have other ethical concerns with regard to PETA. For example, I've seen film footage of the horrible deaths experienced by animals that contract foot-and-mouth disease. I also know the devastating toll that this terrible disease brought to the people of Great Britain. So when Ms. Newkirk says, "I openly hope it will come here," I can't for the life of me figure out how that's in the best interests of anyone. In fact, a US congressman, Representative Scott McInnis of Colorado, recently called into question whether PETA should even continue to be granted status as a tax-exempt organization. His concerns centered around continuing rumors that PETA has supported the Earth Liberation Front (ELF) and the Animal Liberation Front (ALF), which the FBI considers to be among the most significant of the domestic terrorist organizations operating in the US today. Certainly PETA condones organizations like ALF. "I will be the last person to condemn ALF," says Ms. Newkirk. She hasn't ever admitted to torching a building filled with animals herself, but she has said, "If I had more guts, I'd light a match."

Despite all of this, I have a soft spot in my heart for them. The reason is simple enough: Sometimes they're right. Sometimes they're right even when they attack me. Not usually, but at least sometimes.

The first time I ever had a run-in with PETA was over the importation of the rhino we captured in Skukuza. In addition to Cynde and John Barnes, the couple we almost accidentally killed on a previous trip to Africa, Myrta Pulliam, and the others, we also brought along John Stehr, a reporter from a local TV station. John did a wonderful series on the rhino but didn't do us much of a favor when he called PETA's national headquarters to get a comment from them about whether they would consider this the kind of thing that really is good for the species. I could have told John what they were going to say. They don't really care about the good of the rhino as a species to begin with.

So, PETA put the Indianapolis Zoo on their Web site and invited their members to e-mail us, the management of Kruger National Park, and the president of South Africa with requests to stop any importation. They claimed that we were ripping babies from their mothers and using the rhino for economic gain. Both claims were patently absurd, but we got hundreds of letters, most of which simply parroted PETA's short and remarkably inaccurate call to action. I answered every single one and, since most of the letters were more or less identical, I replied to all of them, making the same seven points. They follow verbatim:

> 1. Through the remarkable conservation achievements of South Africa, the white rhino has been brought back from the brink of extinction to numbers plentiful enough for the species to be downlisted from endan-

gered to threatened, which allows for the importation of this species. While numbers continue to grow in the wild, there are no guarantees that this population will continue to be protected and preserved over the next one hundred years. Captive breeding programs are an important insurance policy for all species, especially those whose limited numbers warrant a diversified conservation strategy. By developing self-sustaining populations of rhinos in other parts of the world, the species will continue if some catastrophic event would severely damage or eradicate a population.

2. The zoo chose to acquire its rhino from Kruger National Park in South Africa. This haven for threatened and endangered native wildlife has been a major factor in a fairly dramatic increase in the population of white rhinos over the last several years—from only a few hundred specimens to a current population estimated at 1,800. In fact, their success means that current numbers of white rhinos must be managed to avoid overpopulation, as the park is limited in size and the animals it can support. Overconsumption of plant species due to overcrowding could modify the habitat, adversely affecting many animals, not just rhino. Too many of any species could lead to endemic disease, and even starvation.

3. Kruger National Park gains much-needed revenue through the generous contributions of zoos in exchange for animals such as rhino. This revenue is crucial for the continuation of its programs, for the park itself, and for the protection of the animals that reside there.

4. Following the history-making births of the first and second African elephants in the world to be conceived by artificial insemination, the zoo's growing elephant herd needed larger quarters and the long-term ability to house a bull elephant. To solve this problem, the zoo designed and is in the process of constructing a completely new exhibit. When this is completed, the former elephant exhibit will become available for the exhibition of another large, charismatic species.

5. The zoo has decided that the species best suited for acquisition is the white rhino. All rhinoceros species are threatened or endangered, and research into the reproductive physiology and behaviors of these animals would provide invaluable information for scientists and biologists dealing with animals in both human care and in the wild. That research may lead the zoo to perfect an artificial insemination program for rhinos that is similar to the one devised, and now proven to be successful, for the African elephant. This program is likely to prove essential to the future of captive animal management programs and perhaps ultimately the survival of the species.

6. The zoo acquires a fascinating species for exhibition, research, and

educational purposes. By bringing these particular animals to Indian-apolis, the zoo exemplifies its mission of fostering in its visitors a sense of stewardship for the plants and animals of the natural world. While life for these animals in human care will be different from life in the wild, it will not necessarily be a negative experience. Once adjusted, the white rhinos can look forward to long lives with an abundance of food, top-notch veterinary care, an absence of predation, and the companion-ship of herd mates.

7. It is our belief that when we can act, it is unethical to permit the extinction of any species.

If you read all of that, you'd realize that PETA was wrong about our im-portation of rhino. But in recent days they've been organizing a similar cam-paign, designed to make the Saint Louis Zoo do something about a chimp named Edith. And we *should* do something about Edith.

Let me tell you Edith's story.

Remember when I told you about the "monkey shows" that the Saint Louis Zoo presented back in the days when George Vierheller was the director? Well, back then a lot of chimps passed through the Saint Louis Zoo. Most of them were males. The zoo didn't want females for the simple reason that, when they went into estrus, the chimps became quite a bit more interested in sex than they were in the show. A few females were in the show, but only when they were youngsters. As they approached adolescence, they were sent away. Edith was one of the few chimps born at the Saint Louis Zoo back in Vierheller's day. In fact, only four chimps were born at the zoo during the era of the shows. The reason is pretty obvious—very few females means very few babies.

Edith was born here in 1964. After a few years, she was sent to the beauti-ful new zoo in Louisville. Later she moved to Miami's zoo, and from there she was sold to a dealer, with the permission of the Louisville Zoo. Edith wound up in an unaccredited facility in Texas, and that's where, forty years later, an undercover operative for PETA found her. The job of the undercover opera-tive was to pose as a regular employee in order to document the condition of the animals. Her actual job was to keep Edith's enclosure clean. She took photos of Edith in conditions that were clearly unsanitary and then placed the photos on PETA's Web site. We should probably take it as an article of faith that, even though it was *her job* to keep the pen clean, she didn't stage the photos. At least, I believe her.

PETA recently organized a campaign, asking the Saint Louis Zoo to try and get Edith back.

Now I have to emphasize that the zoo didn't do anything wrong by sending Edith to another zoo, according to the standards of the day. I was eleven years old then. Values and attitudes were different. This was brought home to me on a recent trip that I took to one of our zoo's field sites. I was in Peru, with Mike Macek, our curator of birds, and a few friends of ours visiting a place called Punta San Juan, a tiny village on the coast some seven and a half hours south of Lima by car. This is one of the few parts of the world where the ocean meets the desert. The rain, coming from the east, hits the heights of the Andes Mountains, and all of the moisture falls from the air. On the west side of the Andes, toward the Pacific coast, the land is a harsh, almost lunar, desert landscape.

Punta San Juan is a mining town, built by an American company and then sold to the Chinese (not just the mine, but the whole town with it). People still live there, and the mine is still active, but most of the homes built by the Americans sit empty in the hot desert sun. The town is located near the cliffs, where the Pacific Ocean crashes constantly against the shore, sending plumes of water upwards of one hundred feet in the air. Nothing grows there, but the coastline itself is rich with marine life. The reason for this abundance is that the continental shelf at Punta San Juan is almost nonexistent. This allows the rich, cold waters of the Pacific's Humboldt Current to come very close to the shore, providing food for birds and mammals alike.

Our zoo, in cooperation with Chicago's Brookfield Zoo and the Philadelphia Zoo, runs a tiny marine preserve just outside the town. The reserve itself is a peninsula about a third the size of the Saint Louis Zoo—maybe thirty or so acres—and it is bounded by a cement wall that stretches from shore to shore that keeps predators (in this case, humans and foxes) from coming out on the magnificent point of land that makes up the actual preserve. The reserve is home to tens of thousands of shorebirds, including the Inca tern, cormorants, and great flocks of pelicans, as well as fur seals and sea lions, which bask in the sun along the rocky beaches below the high cliffs. This tiny little preserve is also home to about half of Peru's entire population of Humboldt penguins, which are, along with the Galapagos penguin, the most endangered of the seventeen different penguin species.

The foxes go after the penguins that nest along the shore, but the humans are after something of greater economic value. They come to poach not animals but animal excrement. They're coming to steal poop. Bird guano is an incredibly rich fertilizer, and a bag of it is quite valuable. The problem is that penguins and other birds build their nests in the deep, rich accumulations of guano. If the theft went unabated, the birds would lose their most important breeding site in the entire country.

Anyway, there isn't much to the research station. There's no running water and no electricity, but there is one of the most stunning views of one of the most beautiful coasts in the world. Imagine Big Sur times two, and you begin to get the idea. Likewise, there's not much to the town—a few little shops, a pretty good restaurant, and a small market. Of course, the selection in any of the shops is nothing like Americans are used to. Instead of, say, one hundred-odd brands of cigarettes, there might only be four or five.

I happened to look at the cigarettes and was surprised to see that one brand was Lucky Strikes—the brand my dad used to smoke when I was a child. I hadn't seen Luckies in years, but the packaging looked just the same. I found myself smiling at the recollection of a game we used to play as children. If we were walking down the street and saw a crumpled pack of Luckies that had been tossed away, we would run for the pack, trying to be the first person to step on it. Whoever was first shouted, "Lucky Strikes and no strikes back" and got to punch everyone else on the arm. My recollection is that we punched each other a lot.

When I was eleven years old, it wasn't uncommon to see trash everywhere. Certainly there were enough crumpled packs of Luckies to make for a never-ending game and, at age eleven, I didn't think much about it. I remember quite vividly the day I did begin to think about it, though.

We were living outside of Buffalo, New York, at the time, and our neighbor was a man named Dr. Ken Stewart. Dr. Stewart was one of the early limnologists, a scientist who studies freshwater lakes and streams. His specialty was the Finger Lakes region of New York, and to this day, he is regarded as the man who developed the most comprehensive data set ever for those lakes. He had young children, but they were too young to go with him on his field trips, which lasted all day. He would travel by car to his field sites, go out in a small boat to collect samples and take measurements, and then spend the evening hours back in his lab, analyzing the results. Since he liked the company, he used to take me, and later my brother Jim, along with him.

On my first trip, we stopped in the early morning to buy some donuts to eat in the car. Back then they put a few little pieces of waxed paper in the donut bag, so you could eat them without getting your fingers sticky. We were driving down the road, chomping away, and when Dr. Stewart finished his donut, he casually tossed the waxed paper out the window, and it began to blow across a newly plowed field. Suddenly Dr. Stewart pulled over on the shoulder, opened the door, and began to run frantically after the blowing paper. I thought he had gone nuts. After a few minutes he returned to the car, winded but triumphant, holding the waxed paper. I was, of course, looking at him as if he had lost his mind. He gently explained that he had come to think it was wrong to throw out trash, even in the countryside, and that he was making a conscious attempt never to do it again. Within a few short years, many, if not most of us, would come to believe the same thing.

But back to Edith. In 1964 we sent chimps away for reasons of convenience. In 2004 we don't. We would never break up a social group unless it was part of breeding recommendation by the chimp SSP. In fact, our entire chimpanzee program is built around the idea of building social groups. Our program specializes in building foster families, as we saw earlier with the story of Tammy. With the benefit of hindsight, we can say that letting Edith go was wrong and that we have an obligation to do something about it. The question is, "What should we do?"

The problem is that for Species Survival Plans to work, we need to cooperate as institutions and guarantee to one another that we will never send an animal that is genetically surplus or one that is recommended for breeding to a nonaccredited facility *unless* we can be certain that the facility provides adequately for the welfare of the individual. In other words, while the welfare of the species is vital, it is not more important than the welfare of an individual.

In Edith's case, this means that the Louisville Zoo is the one that should take responsibility. If it sounds like I'm trying to pass the buck, I'm not. As a matter of fact, while only four chimps that were born here ever wound up in nonaccredited facilities, at least eighteen chimps that once lived here eventually wound up in nonaccredited facilities. My standard imposes a much more difficult task for our zoo. If we just took responsibility for animals *born* here decades ago, we would have a problem with four chimps (including Edith). If we took responsibility for animals that weren't born here, but *lived* here for a time and are now in nonaccredited facilities, we have a problem that's over four times as large.

To further complicate things, back in the 1960s accreditation didn't even exist! It may seem a little hard on ourselves to impose today's values on yesterday's actions, but I think we must. The Saint Louis Zoo, along with the other zoos that bred or held chimps over the years, has an obligation to those animals—Edith and all the others. And we will find any remaining chimps (an animal that can easily live forty years in human care), and we will ensure that they are living comfortable, rewarding lives. We may have to bring them back to St. Louis, find another zoo that will take them, or find a sanctuary that we trust, but we must track all of them down and we must do it as quickly as we can. Both my head and my heart tell me that this is the right thing to do.

This brings me to my central problem with PETA. They misrepresent the truth, squander precious resources on defending animal activists accused of crimes, euthanize innocent animals because it is inconvenient to keep them alive, and spend, as near as I can tell, absolutely nothing to help animals in the wild. Very often, though, their hearts are in the right place.

Their methods, on the other hand, are worse than self-serving. For example, had they come to us, quietly, and said, "there's a chimp that is living in less than desirable conditions, and we'll give you two weeks to do something about it. If you don't we'll go public," then we would at least have a fighting chance. Instead, they went public first. That puts the man who has Edith in the position of admitting that he is at fault, something that no one ever wants to do. The upshot is that the person who has Edith has publicly stated that he will never allow Edith to leave. Had PETA come quietly to us, we might have gotten Edith out. After all, she's too old to have babies and is probably of virtually no value to the current owner.

But PETA figures to win either way. If we're successful in getting the chimp out, they can take the credit. If, thanks to their precipitous actions, we're not successful, then they can blame the zoo. Of course, while they may win, Edith is the big loser. It's a smart PR move for them. I mean, that's using your head, at least from PETA's point of view.

It's too bad that, somewhere along the way, their head got disconnected from their heart.

Zoos of the Future

The world around us is changing fast and, when it comes to wildlife, it is not changing for the better. The IUCN publishes the international list of threatened and endangered species (called the Red List), and a quick look at their list will tell you this: 18 percent of the world's remaining mammals and 11 percent of the world's remaining birds are threatened with extinction. As Bill Conway recently summarized, "Almost all large animals are in trouble; storks and cranes, pythons and crocodiles, great apes (in fact, most of the primates), elephants and rhinoceroses. Ninety percent of black rhinoceroses have been killed in the past eighteen years and one-third of the world's 266 turtle species are now threatened with extinction." Amphibians are disappearing worldwide. Add to that reports of acid rain, ozone depletion, global warming, the destruction of the world's rain forests, and phytoplankton blooms and coral bleaching in our oceans, and you get the picture of a world on the precipice of environmental disaster.

In his essay, "The Changing Role of Zoos in the 21st Century," Bill Conway quotes Jack Welch, CEO of General Electric, who once said, "When the rate of change on the outside exceeds the rate of change on the inside, the end is in sight." The problem, according to Conway, is that the "outside" world of wildlife and nature, which the world's zoos represent to millions of visitors every year, is clearly changing faster than the "internal" zoo response. In other words, zoos have got to start doing things differently. And they have to start now. We have been overtaken by the magnitude and speed of extinction. If we do not respond, the zoo of the future will be little more than a living museum.

In 2004 a small group of zoo professionals from around the world gathered in London to think about precisely this problem, "How can zoos transform themselves so that they'll be able to respond, in a fundamental way, to massive global extinctions?" The conclusion of this conference was that zoos must redefine themselves in a completely different way—in the words of one zoo professional, they must become *in situ* conservation organizations. This means that the fundamental role of zoos, their reason for being, must become the preservation of animals in the wild. Much like Conservation International (CI) and the World Wildlife Fund, our zoos must become protectors of wildlife and wild places. But there are fundamental differences between CI and WWF and zoos. Zoos have a whole different set of strengths and assets than international conservation organizations (of course, they also have some problems that the major international organizations don't have).

Zoos have several things that make them unique. First, zoo professionals are the world's experts on breeding small populations of endangered species. No other class of research or conservation organizations, universities, in fact, nobody else, has that skill set. Second, we have living things. We have an incredible variety of living things—everything from anteaters to amphibians, birds to butterflies, conger eels to capybara—I could go through the entire alphabet, but you get the idea. That is a resource no one else has. Third, zoos are already used to collaboration. We cooperate in the management of all of the animals that are in Species Survival Plans (SSPs), we collaborate on research projects, and we collaborate on field efforts like those of the Madagascar Fauna Group. Fourth, we have visitors. WWF and CI don't get the more than three million visitors a year that the Saint Louis Zoo does. That is a huge, albeit largely untapped, resource.

So these are, I think, our major strengths. We hold over ten thousand different species of animals from all parts of the world, many of them rare or endangered; we host more visitors annually than all major professional sports combined; and we have unique expertise in the management of animals, unique capabilities for research on exotic species, and a high potential for developing conservation collaborations. In the end, I believe that zoos hold extraordinary promise, perhaps the greatest promise of any type of conservation organization, for preserving wild things in wild places, providing a safety net for charismatic species in danger of extinction, and mobilizing the interests and passion of the general public for worldwide conservation.

The first thing we must do, as a consequence, is to develop a list of our priorities. Both CI and WWF have already done so. They looked around the world and asked, essentially, "Where are the areas of greatest biodiversity, the areas richest in life forms, and which of those areas are in the greatest need of protection?" We could, I suppose, just adopt either of those two lists, but that wouldn't allow us to take maximum advantage of our strengths. Let me give you one example. The largest animal to go extinct in the wild in the last ten years is the scimitar-horned oryx. It was once found in abundance in the broad band of arid and semiarid lands south of the Sahara. This very special arid zone extends east from the Atlantic Ocean all the way to the Horn of Africa and touches some seventeen different countries. Technically called the Sahelo-Saharan region, it is not found in either CI's or WWF's list of hot-spots. Why? Well, because there just aren't that many different species of plants and animals in this vast arid land. As a consequence, conservation of the scimitar-horned oryx, along with the numerous other species of gazelles (like the magnificent addax), predators (like cheetahs, foxes, and a host of others), birds, reptiles, and even amphibians, goes largely neglected, and many of those species are almost extinct. Zoos can, however, easily breed most of the animals that are native to the Sahelo-Saharan and on the threatened or endangered list. There are also thousands upon thousands of square miles of habitat available for reintroduction of these animals. So while this region may not be as biologically diverse as, say, the rain forests of Madagascar, it is home to many wonderful creatures, it could be easily repopulated, and it most certainly deserves far more attention than the international conservation community has given. I would think that a region like this would be pretty high on our list, even if it is not high on CI's or WWF's. We can really do something about the Sahelo-Saharan region. In all likelihood, we won't.

So, first, zoos need to develop their own conservation priorities. In many cases they will overlap perfectly with the priorities of other conservation organizations, but in some cases, they will not. Second, we must reorganize ourselves internally to meet the prioritized need. This means that our collections and thus our breeding programs need to relate to our priorities. I've already mentioned that many zoos hold common zebras, even though the rarest of the three species of zebras, the Grevy's zebra, is rapidly becoming extinct. That just doesn't make any sense. Zoos can be an important safety net for the Grevy's, and if they become extinct in the wild or, almost as bad, their populations become so small that they are genetically compromised, we can reintroduce animals back to the wild.

In other words, zoos have to achieve what we might call "integrated seamless conservation" from our zoo out to the wild—from inside our fence, back out to the field. The Madagascar Fauna Group is often used as an example. We have bred lemurs in zoos, returned them to our zoo in Madagascar, allowed them to learn how to survive in the wild, and then released them back to the wild. That, in a nutshell, is integrated conservation.

This leads me to the third thing that must happen. Zoos cannot work alone. We have to partner with organizations like CI and WWF, with universities, and with a wide variety of local conservation organizations, governmental agencies—indeed, with a vast arena of potential partners—if we are to be successful. Again, the MFG provides a pretty good example. We work with Duke University and SUNY–Stony Brook very closely. Conservation International is a wonderful partner, as is the Wildlife Conservation Society. But we also work with the government of Madagascar and with two major universities in Madagascar. Nothing happens without that high degree of co-operation and collaboration. Nothing.

Finally, zoos must lead the way in pushing for a massive shift in social and political support for conservation. This means not only that we have to convince our millions upon millions of visitors that they should be interested and invested in these efforts, but also that we need to work with our government and other governments to push this agenda. Most zoo directors and zoo trustees are scared to death of doing so, but we simply have to. We must advocate for wild things and wild places, and we must use our considerable political muscle to do so. Zoo directors are getting pretty good at lobbying, but our boards are rarely, if ever, called on to help. That must change.

It used to be, decades ago, that we used boards primarily, in fact almost exclusively, as fund-raisers. That accounts for the customary "three Gs" of board involvement (*get* us money, *give* us money, or *get* the heck out of here). But as discussed earlier, now it's more common to think of boards in terms of the "three Ws"—we want your *wealth,* yes, but we also want your *wisdom* and we want you to *work* hard for our institutions. Perhaps now we should talk about the "three Is"—we certainly still want boards to generate *income,* but we also want them for *inspiration* and, most importantly in the emerging world of zoo conservation, we want their *influence.* The battle for saving wild things in wild places will not be fought in the field as much as it will be waged on the political front.

In short, we have to change in two profound ways. We have to change our basic orientation in order to become full and complete field conservation organizations, and we have to change in terms of what we communicate to our audience and how we utilize our vast audience to understand and influence both the social and political directions that our society takes.

Of the two, I am confident that we know the most about how to reorganize ourselves in order to become *in situ* conservation organizations. As an example, let me tell you how we are doing it here in St. Louis.

When I first came to the Saint Louis Zoo, I would have described our conservation and research efforts as a mile and a half wide and a half-inch deep. In the three years prior to my arrival, we were engaged in over one hundred different conservation efforts in over twenty different countries. Funding for these efforts came from our general operating budget, and most of the projects we funded received small amounts of money, anywhere from $1,000 to $10,000. Our research efforts (what we call *ex situ* work, or work on animals in the zoo) were not fully integrated with our conservation efforts, but they were very important all the same. There is an enormous amount of information that can only be gained by having animals in zoos. For example, earlier we saw how most of what science knows about the hormonal basis of elephant reproduction was learned in zoos (much of it spearheaded by Indianapolis and the National Zoo, but data come from virtually every zoo in America that keeps elephants). That knowledge has tremendous implications for how we manage populations of elephants in the wild. Make no mistake about it; those populations *are* managed. Although they're not as intensively managed as rhino (which pretty much require an average of one armed guard per animal to survive), we still have to protect elephants and manage their population size, even in national parks. Our forte, though, is precisely that. We understand the genetic dynamics of managing small populations better than any other type of organization. We understand effective intervention and wildlife medicine better than any other type of organization. We do more research on things like animal nutrition than any other organization, and we have a unique capacity for doing controlled behavioral studies that no one else can match. Finally, we have representatives from a larger number of species to work with than any other type of organization—even if you added all of the species held in all of the universities in the world together, it wouldn't come anywhere close to the ten thousand different species represented in ISIS, the world's zoo and aquarium database.

The first thing we did at the Saint Louis Zoo was to say, in effect, let's cut our conservation activities way down, so we are active in much fewer places, but at the same time, let's dramatically increase the amount of total funding we're putting into those places. This should mean that, while we're no longer making a little bit of difference in a lot of places, we are making a profound and lasting difference in a few places. Then let's work hard to ensure that our research program is focused on helping those conservation efforts, our animal collection becomes more focused on those efforts, and our exhibits and educational programs begin to get the message out about what we're doing and why it's so important.

To this end, we decided that we should package our smaller number of larger projects together into a single entity, something that (with the help of some wonderful outside marketing experts) we decided to call the WildCare Institute. The WildCare Institute is dedicated to creating a sustainable future for wildlife *and* for people around the world. The idea was to make each curator and staff scientist in the zoo, twelve people in all, pick one area and focus all of their attention on that area. We decided that we would require each person to develop working collaborations with other organizations and institutions, and focus his or her efforts on habitat protection, community development, and what we call "capacity building," or training. The idea here is that nothing happens in conservation unless the people living in the areas where we're working can see a direct benefit to themselves. Likewise, we won't be successful unless those same people can carry on the work of conservation by themselves. That's why we choose to emphasize community development and capacity building—the training of local folks to work in conservation so that we can eventually move on to other parts of the world on our priority list.

Fortunately, we had a wonderful person already on staff to head this major new effort. Dr. Eric Miller, a vet who trained at Ohio State University, had started his career at the Saint Louis Zoo as a veterinary resident back in 1981. He joined the zoo staff as a vet in 1983 and has been the zoo's director of animal health and conservation since 1993. In effect, he grew up here. Eric defined the twelve areas where we should be working in conjunction with the staff, and then he approached the Saint Louis Zoo Friends with an idea for securing a dedicated source of funding. While I was away celebrating my fiftieth birthday (hiking in the Kruger with Patrick Boddam-Whetham), Eric was quietly negotiating a big-time birthday present from the Zoo Friends—a nineteen-million-dollar gift that would launch the new institute. He also got our Board of Trustees and our Zoo Friends to agree to give all of the revenues from our beautiful new Mary Ann Lee Conservation Carousel to the institute every year. All of this led me to believe that I should take more vacations (or, alternatively, just keep celebrating my fiftieth birthday over and over again).

The twelve areas that Eric and the staff picked were all areas in which, to one degree or another, we were already active and invested. Two efforts are focused here in Missouri and the other ten are scattered around the globe.

Of the two Missouri centers, we've already heard a bit about the first. This center focuses on the American burying beetle, that charming little invertebrate that so carefully rears its young (and eats them, too, if there're too many). Jane Stevens, our curator of invertebrates, picked the beetle as a flagship species for a variety of reasons. First, they are inherently fascinating. Second, the Roger Williams Park Zoo, in Rhode Island, had already developed the protocols for rearing them in zoos and had even started releasing them back to the wild. Third, they were once plentiful in Missouri, but despite many field trips, Jane has been completely unsuccessful in locating any beetles anywhere in the state. Fourth, she thought that she could find a variety of partners in addition to Roger Williams Park, partners that could share expertise. For example, Eric's alma mater, Ohio State, had beetles that it could transfer to us to begin a breeding program here at the zoo, and The Wilds, a huge landscape-scale reserve operated in conjunction with the Columbus Zoo (also in Ohio), was already releasing beetles and studying their survival. Jane's center, although in its infancy, is already breeding beetles and has scheduled a major PHVA (or Population Habitat and Viability Assessment) workshop with the ever-helpful CBSG. Like the PHVA for the black-footed ferret and golden lion tamarin, this PHVA holds the promise of developing an integrated plan for the recovery of the beetles in the wild. It will probably take the better part of a year, but by the end of 2006, Jane and the other participants in this project may well have a plan in place that integrates the efforts of the Missouri Department of Conservation, the US Fish and Wildlife Service, universities, and other conservation organizations, such as The Nature Conservancy, in a plan that we hope will yield the reintroduction of established populations of beetles back into the remaining wild places of our state.

The second Missouri center is sponsored by our general curator, Ron Goellner (the same Ron who found the escaped cobra some twenty years ago and who went on to found the SSP for *Partula* snails here in America). Ron loves creeping, wiggly things, and he's always had a fascination for a very peculiar amphibian called a hellbender. I know that sounds like a name for a rugby team or a motorcycle gang, and unfortunately, nobody knows why they came to be called hellbenders, but we do know that our state is the only place in America where both subspecies of hellbenders can be found.

Hellbenders look like a cross between a frog and a Shar-Pei. They are almost two feet long when fully grown, and they live under rocks in Missouri's fast-flowing streams. They are brown and wrinkled (having more skin means that they can absorb more oxygen from the water), and they have beady little eyes. They live for a long time—upwards of fifty years—and even people who float our fast streams rarely know they are even there.

Some ten years ago, scientists noticed that hellbender populations were beginning to decline in Missouri. At first we weren't too worried, but now we are deeply concerned. One of the first things Ron did was to fund Dr. Max Nickerson, the field researcher who first started to census the hellbenders in the wild back in 1969, to conduct an updated census. What he found was that in stretches of river where he last counted hundreds of hellbenders per mile, he was now finding only a few animals, and many of those had deformed or missing limbs. In fact, our recent survey shows about a 95 percent decline in their numbers. Knowing the extent of the problem is important, but it is even more important to understand *why* they're in decline. Ron has offered several theories. Perhaps the rise in canoeing and floating on those rivers has disturbed the habitat. Many of the rivers are pretty shallow, and the fry, or young hellbender salamanders, might be getting crushed as canoeists push their boats over the shallow gravel bars. Or perhaps the young are being eaten by an introduced species. The rivers they live in are perfect for trout, and trout have been introduced in most of the rivers where hellbenders were historically plentiful. The most interesting theory is biochemical. When the Department of Conservation started bringing us hellbenders to examine, our reproductive physiologists discovered that although the females were chock full of nice big eggs, the males had very few sperm, and many of their sperm were deformed or abnormal. Perhaps, Ron reasoned, growth hormones from cattle feed were leaching into the rivers and "feminizing" the males, or perhaps more likely, agricultural pesticides or insecticides were leaching into the water systems and having much the same effect. I thought that was pretty silly until I read that rural Missouri men had lower sperm counts and poorer sperm quality than urban men. Perhaps, one wonders, whatever is affecting the hellbenders is affecting us. Ron is now breeding hellbenders here at the zoo, and his center, like Jane's, is planning a PHVA that will, we hope, yield a plan for the conservation of the hellbenders that remain in the wild. Having two Missouri species, hellbenders and beetles, gives us a nice combination of aquatic and terrestrial and, we hope, opens up funding possibilities for Missourians who are interested in doing important conservation work close to home. It also gives us two species that we might call "bellwether species." Hellbenders and burying beetles might be compared to canaries in a coal mine. Sensitive to even subtle changes in the environment, they might tell us that something is very wrong with our environment before whatever it is begins to affect us (although if I was a rural Missouri man, I might begin to wonder if I was already profoundly changed in, let's say, a most personal way).

We have already talked about some of our work in other parts of the world. Ingrid Porton's center is based on the work of the Madagascar Fauna Group (which is now headquartered here at the Saint Louis Zoo). In many ways, this is a model center. It is highly collaborative, it actively works to preserve critical habitat by operating a major reserve, it provides employment for thirty-two Malagasy people by operating a zoo and staffing the reserve, and it actively trains those employees to take management positions. In fact, all of our managers except two are native Malagasy who have been trained by the MFG. Some day, that number will be zero, an indication that our training efforts have paid off.

Cheri Asa, the reproductive physiologist who is also working with the Channel Island fox, picked Bosawas, Nicaragua, as her center. Our zoo has been actively cooperating with Mesoamerican zoos for quite some time, and so it was no surprise when The Nature Conservancy (TNC) approached us with a fascinating opportunity. TNC had been working in a place called the Bosawas Biosphere Reserve in northeastern Nicaragua, the largest tract of tropical moist forest north of the Amazon Basin. This vast tract of land is home to three thousand indigenous Miskito and Mayangna people, who live largely as their ancestors have for generations. They grow a few crops and hunt in their forest, but they have never caused any significant habitat disturbance. In fact, 90 percent of the reserve is still considered primary forest, having never been logged. The natives' traditional way of life is threatened, however, by mestizo (mixed-race) colonists, who are sneaking into the reserve and practicing a decidedly environmentally unfriendly form of agriculture—slash-and-burn. The Nature Conservancy is working with the indigenous people to help them gain title to their lands, thus keeping mestizos out of the reserve. To get clear title, they must have a land-use plan that demonstrates that their traditional lifeways are, indeed, sustainable. They asked our zoo to come and survey all of the animals in the forest with special reference to the animals they traditionally hunt. Second, they asked us to do what anthropologists call "cooking-pot studies." These studies involve recording everything about the animals they hunt, the species, the sex, the size, and so on. The women are most helpful in this regard, because they have no reason to misrepresent what they cooked for dinner. Men, on the other hand, have a tendency to exaggerate a little (witness me coming back from a fishing trip and telling my wife, "I caught a fish *this big!*"). If we can prove to the government that the indigenous people can effectively manage their natural resources, TNC is confident that they can help gain clear title to this incredibly rich and important habitat. As in Madagascar, most of the people involved in this effort are indigenous peoples. Right now, we employ over thirty natives, whom we have trained to do wildlife census work and cooking-pot studies. They work under Drs. Dan Polisar and Dan Griffith, of the zoo staff, and as in Madagascar, we collaborate with a number of universities as well. Penn State is one of our most active partners, but Saint Louis University and Idaho State also provide vital resources, as does the University of Missouri–St. Louis through its International Center for Tropical Ecology, along with another important partner, the Missouri Botanical Garden. Recently our program expanded to include the work of Louise Bradshaw, the zoo's director of education. Louise has been down to work with teachers in the little schools that dot the reserve, developing curriculum on the natural environment in the native languages.

Another center that is very well developed is Eric's own center, which focuses on avian (or bird) health in the Galapagos Islands. I guess I shouldn't say it's Eric's alone. We have a joint professorship with the University of Missouri–St. Louis, sponsored by a man named Des Lee (he's the husband of Mary Ann Lee, the woman who put our Conservation Carousel in the zoo). Des is a huge believer in collaborations and has funded endowed professorships between the university and a variety of cultural institutions in the city. I can, modesty aside, claim that ours is the best. Our joint professor, Patty Parker, is the chair of the biology department and is a world authority on population genetics. Much of her effort (and that of many of her graduate students) focuses on the Galapagos, one of the most fascinating places left on our planet.

Unlike the Hawaiian Islands, where sixty of the original eighty-eight species of birds found only in Hawaii are now extinct, the Galapagos has not lost a single bird species in recorded history. At least two species, the mangrove finch and the Galapagos penguin, are teetering on the edge of extinction, but so far at least, no species endemic to the Galapagos is gone. The Charles Darwin Research Station, the Zoological Society of London, and the

Galapagos National Parks have partnered with the zoo in an effort to ensure that no bird diseases (the most likely cause of a future extinction) make their way to the islands. To this end, our zoo was the first conservation organization in the world to have a full-time field pathologist. Our pathologist works closely with Patty and her students, doing what is essentially a biomedical study. Every time Patty takes a blood sample from a bird for genetic analysis, we come along behind her and examine the sample for the presence of any blood-borne disease and look at the bird for things called ectoparasites. Our biggest fear is that, eventually, our old nemesis West Nile virus will make its way to the islands. It will happen. The only questions are when, and will we be ready when it does?

Some of the centers are not as well formed as the ones in Madagascar, Nicaragua, and the Galapagos. In fact, I had the fabulous opportunity to be involved in the formation of a center dedicated to the conservation of Grevy's zebra in Kenya and Ethiopia. The center came into existence as a direct result of the charisma and drive of Martha Fischer, our curator of ungulates. We met Martha earlier in this book; at first glance, she doesn't look like the stereotypic notion of a field scientist. A statuesque blonde, it's a little hard to imagine her riding on horseback in search of the Ethiopian wolf in the wilds of the Bale Mountains or traveling through war-torn areas of that country counting zebra. But she has an abiding love of the people and the animals in the Horn of Africa and a singular ability to impart that passion to others.

Martha's special love is the Grevy's zebra, a species of zebra that has suffered a decline in numbers of somewhere between 86 and 89 percent in Kenya since 1980 and a 90 percent decline in Ethiopia since 1980. Historically found only in these two countries, their total numbers are somewhere between 1,700 and 2,100 remaining animals. Worse, only 0.5 percent live in protected areas.

About 70 percent of the remaining animals are concentrated in a small area spanning some seventy miles between Lewa Wildlife Conservancy, a private reserve owned by the Craig family, and three contiguous government parks, Samburu, Shaba, and Buffalo Springs. The threats to Grevy's are similar no matter where they're found—habitat loss and degradation, competition with domesticated livestock, poaching for meat, and predation. The last threat is interesting, because lions, at least according to our research in Lewa, prefer to eat Grevy's zebras more than they do common zebras. Common zebras form herds that wander the landscape. Grevy's, a species adapted more to the arid lands of northern Kenya and Ethiopia, are territorial. A stallion stakes

A Grevy's *zebra*.
MARTHA FISCHER

out a claim, usually close to a good water source, and then entices mares into his harem. Since they can always be found in roughly the same place, the lions say, in effect, "Room service!" Researchers at Lewa have radio-collared lions, which they track every day. They collect the lion's feces, dry them, and examine the zebra hairs under a microscope. This is how we know that lions have developed a preference for Grevy's over the common zebra. (An amazing amount of field research utilizes fecal samples for one purpose or another!)

The Samburu people who live in the communities between Lewa and the parks, thanks to tireless work by the folks at the Lewa Conservancy, have organized themselves into something called the Northern Rangelands Trust.

Beginning with the community closest to Lewa, a place called Il Ngwesi (which rhymes with "ill and queasy"—I once had a mild case of the flu while staying there) was the first community to begin working with endangered species on their community-owned lands. What they did, with Lewa's help and guidance, was to set part of their tribal lands in reserve as a core conservation area, an area where they would not graze their cattle, sheep, and goats. In a few short years, wildlife began to move into their conservation area, and they monitored that wildlife closely. They started a "scout program," which the Saint Louis Zoo funds (not only in that community, but in all of the communities that constitute the Northern Rangelands Trust). The scout program, run cooperatively by Lewa, Princeton University, and our zoo, trains villagers, both men and women, to survey their community lands for Grevy's. Although the villagers are illiterate, they can use preprinted forms with symbols for the daily weather (the sun, the sun obscured by clouds, and so on), symbols for the vegetation where the animals are sighted, symbols for stallions, mares, and colts, and other observations. They also have a portable GPS unit. After they record their data, they push a button and lock in the coordinates for where they made their observations. This has made our partnership the largest single employer in the region (and the leading employer of women, as well). In addition to the scout program, we monitor Grevy's by radio collaring animals and tracking them, and by performing regular aerial and ground surveys.

Lewa also has taken a lead role in protecting the Grevy's from poaching. Certainly the scout program helps, but Lewa also has well-trained and well-equipped rangers who can respond quickly if poachers wander into the community. Education is another major initiative. Right now, we have four different education programs that reach children in every school in every one of the seven communities that make up the Northern Rangelands Trust. Of course, Lewa also helps these communities to develop income-generating activities other than the employment provided by the scout program. Il Ngwesi is perhaps the furthest along this road, having developed their own ecotourism lodge. Finally, Lewa helps with operational and technical support, building roads, airstrips, and so forth.

While our zoo sponsors the scout programs in all of the villages, we have taken on the community of Kalama (a tribal community about the same size as Il Ngwesi) as our own special project. Our efforts in Kalama have been devoted to bringing a steady supply of water to the villagers. In exchange for setting aside their own core conservation area, we brought water in for their grazing animals *and* brought water into the conservation area for the zebra and other wildlife. The water will begin flowing soon, and within a matter of a few years (we hope), the wildlife will return to their community in the same way that zebras, wild dogs, and even rhinos have come back to Il Ngwesi.

Many zoos are working with us in northern Kenya. The San Diego Zoo just pledged fifty thousand dollars a year for five years to begin working in West Gate, the tribal land immediately adjacent to Kalama. But many other zoos have played an integral role in the rapid development of this project, including Reid Park, the Oklahoma City Zoo, Brevard Zoo, Albuquerque Zoo, the Dallas Zoo, and the Sedgwick County Zoo. Perhaps the best indication of the value of that cooperation comes from one of our scouts, Rikapo Lentiyoo, who says, "There are opportunities that are arising from this project that are good. Some other areas are crying because they destroyed their wildlife and they can't get them unless they buy them. Nkai, the Samburu God, has given us that heart of taking care of wildlife and people are supporting us to do this. Let us come together, work together, and be united." I could never have said it better.

Another center that I was around for from the beginning (sort of) was our center devoted to the preservation of Humboldt penguins at Punta San Juan. I've already described this magical place where fully half of Peru's Humboldts come to breed. The center was actually initiated by the Wildlife Conservation Society many, many years ago, but in 2002 they found they could no longer fund this little tiny reserve. Thanks to the efforts of Dr. Patty McGill at the Brookfield Zoo in Chicago and the kindness of Pete Hoskins at the Philadelphia Zoo, we have formed a partnership that will ensure that we can keep the reserve staffed and running for the years ahead. Interestingly, Myrta Pulliam, whom we met in the chapter on our rhino capture in the Kruger, was our first private donor to this project. Thankfully, Myrta loves penguins even if her heart is in Indianapolis.

The rest of the other centers are really just getting started, but they seem to grow by leaps and bounds. Bill Houston, our assistant general curator, has a number of other colleagues at other zoos, and together they are beginning to work in the Sahelo-Saharan region. We've already read a bit about Jeff Ettling's project with the lovely but deadly Armenian vipers. Mike Macek, our bird curator, is working with the Humboldts and also has a developing center that will, we hope, preserve the highly endangered horned guan, a bizarre-looking bird found in the highland Chiapas region of Mexico. Alice

Seyfried and Mark Wanner, from our Children's Zoo, are working with Roger Williams Park Zoo on a project in New Guinea, and Steve Bircher, our curator of carnivores, is beginning work on a center devoted to the conservation of cheetahs. All of these efforts show great promise, and all of them connect animals that visitors can see here at the zoo with vital conservation efforts in the field. But none of them are ours and ours alone. All of them rely on a host of partners for their ultimate success.

Communicating all that we're doing to our vast audience is, however, another thing entirely. In the introduction to this book, I talked about how zoos need to get to people's hearts before we get to their heads. After all, we've been giving people information about the declining state of our planet for a long time, and I'm not sure that it has made a world of difference. The problem really turns on the fact that because there's so much information out there, it's hard to know what to care about. That's the downside of the information age. We do know a lot; we don't necessarily know what, of what we know, is important.

I think I know how we can get children interested in conservation, but I'm not sure about the best ways to get their parents interested in the actual projects that I've described in this chapter (although I suppose a book like this is at least one place to start). If I could create the ideal environment for children, I think I know what it would look like. It would look a lot like where my son went to elementary school—a place call New City School, here in St. Louis. The learning that happens in New City School is based on the work of a Harvard psychologist named Howard Gardner. He came up with his theories on learning while working at the Boston Veterans Administration Medical Center, where he first began to explore how people lost different mental abilities if they had injuries to different parts of their brains. On the other hand, they also retained other sets of abilities. He called those sets of abilities "intelligences," and he concluded that there was, in effect, more than one kind of intelligence. In fact, Gardner identified eight different kinds of intelligence, and all of us, to one degree or another, have them in different measure. In no particular order, the intelligences that Dr. Gardner identified are, first, *intrapersonal* intelligence—the ability to access one's own emotional life as a means to understand oneself and others. People with this kind of intelligence make wonderful poets, but they make terrible diplomats. Diplomats, on the other hand, are blessed with what Gardner calls *interpersonal* intelligence—the ability to understand people and relationships. Some people have a great deal of what he calls *spatial* intelligence—the ability to perceive the world accurately and to recreate or transform that perception. Architects, for example, tend to have wonderful amounts of spatial intelligence. The fourth intelligence Gardner identified is *bodily-kinesthetic* intelligence—the ability to use the body skillfully and handle objects adroitly. I've always said that Michael Jordan and Michelle Kwan are geniuses more from the neck up than the neck down. Obviously, some people have more *musical* intelligence than others, a certain sensitivity to pitch, melody, rhythm, and tone. I have a brother who is a mathematician. He has an inordinate degree of what Gardner calls *logical-mathematical* intelligence—the ability to handle chains of reasoning and recognize patterns of order. Many people have *linguistic* intelligence—a sensitivity to the meaning and order of words (think about any author other than, say, me). Finally, Gardner recognized the intelligence I esteem the most, *naturalist* intelligence. These are people like Linnaeus or Darwin, or explorers like Lewis and Clark, people who have an innate ability to recognize and classify what they see in the natural world and an affinity for a myriad of living things.

Given these intelligences, if I wanted to connect people in general, and children in particular, with living things, I would create learning environments that relied on their native strengths. For example, for children blessed with musical intelligence, I would create a place where they could create a symphony of nature, where they could hear the sounds of the wild—whales or birds, insects or mammals—and compose their own unique songs. I am not blessed with an extraordinary degree of musical intelligence, but I am acutely aware of the sounds that nature makes, the rhythms of the cicada or the tree frogs, the tones of a whale's song, or the melody of a lark. Children who have bodily-kinesthetic intelligence might not be as interested in that special place (although musicians often seem to have both types of intelligence in abundance). On the other hand, they might be very interested in a place where they could swing like a monkey, climb like a spider, slither like a snake, or leap like a kangaroo. For children with naturalist intelligence, we could have something like the Dallas Zoo's Trading Post. Here children can bring in natural objects like seashells, mounted butterflies, rocks, fossils, snake sheds, or whatever and trade them. Children with this type of intelligence are the ones not with one pet, but a dozen. Their windowsills are lined with their collections. I think they would love a place that would help them connect to living things in the way they do it best. We could certainly imagine all kinds of environments where children could play at science in a way that connected them to living things.

Of course, an environment like this would not necessarily help to impart the more negative messages about what is happening to our planet, but I'm not sure that this is really an issue. In fact, I think that most children probably aren't emotionally equipped to deal with those issues until they reach their earliest teenage years.

For older children and adults, however, we must impart some sense of the problems that are facing wildlife worldwide. We also have to let them know what we are doing about those problems, at least in the major conservation centers where our zoo has focused it efforts. That begins to get very tricky. It is tricky because while, on the one hand, there is an enormous amount of good news—news filled with hope, joy, and passion—there is also a ton of horrible news, news laden with guilt, fear, and shame. Despite our best efforts, Grevy's zebra is still in decline, there are probably somewhere between 200 and 500 addax left in all of the vastness of the Sahelo-Saharan region, and the mangrove finch of the Galapagos Islands only number about 70 living birds. I must confess that I do not know how the zoo of the future will impart these messages to our vast audience, much less how our zoo will tell our visitors about what we're doing and why it is important.

In the end, that may be the single largest unknown as we look ahead. We simply do not understand how to get our audience to care.

But I do know this: If we cannot get them to care, deeply and passionately, about what we do, then zoos have no future. And if we have no future, then the rare and wonderful species we have elected to save have no future, either. Perhaps it is said best in Ecclesiastes: "For the fate of the men and the fate of the animals is the same, as dies one, so dies the other." If it is true for all individuals, then surely it is true for all species. If it is true for one life, then surely it is true for all of life. Our fate and the fate of other animals is intertwined.

18

Sailing with Noah

I grew up in a devoutly religious family. My grandfather was a Methodist minister who preached in western Pennsylvania and along the southern tier of New York—the counties that border the northern edge of Pennsylvania. He had four sons, and my father was his second. I don't think that any minister preaching in those rural counties in the 1920s was exactly well-off, but my father's earliest memories were good ones. My grandmother died when the boys were quite young, however, and from then on their lives took a decided turn for the worse. My grandfather died from an infection, and the brothers, then ages four through twelve, all wound up in a Methodist orphanage near Jamestown, New York, called Gerry Homes.

I have visited there. Dad still had friends in the small town of Gerry, and much of the orphanage was still standing in the 1960s. We visited the orphanage not because Dad had particularly fond memories of the place but rather because the woman who I had always referred to as my grandmother lived in a part of the orphanage that had been converted into a rest home. She had been the matron of Dad's ward and would live out the rest of her life where she had worked as a younger woman. I guess she was the only mother my dad really knew, and the Methodist home was the only home he ever had. You can't get much more Methodist than that.

Both my parents went to Greenville College, a Methodist school, on the GI Bill, and that's where they met. When I was young they were still pretty strict as Methodists go, perhaps because they were raised as Free Methodists (as opposed to plain old Methodists; the use of the word *free* actually makes for a wonderful oxymoron—they were pretty restrained). In my parents' day, "good" (conservative) Methodists did not wear jewelry. Instead of an engagement ring, a young lady would receive a watch. A watch was okay because it was utilitarian and had the added benefit of ensuring that you had no excuse for being late to church. Makeup was frowned upon, dancing was considered to be a little licentious, and card-playing, while not a sin, was at best a frivolity.

Not surprisingly, when I was young we attended church just about every Sunday. I was encouraged to read anything and everything, but what I remember reading the most (along with the *World Book Encyclopedia*) was *The Bible Story*. There must have been ten volumes in the set; it basically took stories from the Bible, illustrated them with magnificent pictures, and retold them in a fashion that a young child could read and understand. To this day I can conjure up the image of Daniel in the lion's den and Noah building his ark.

As a child, I took a profoundly literal interpretive slant on those stories. No one told me at the age of nine that it was okay to think of those stories as metaphors or abstract lessons. Indeed, I was from a "God said it, I believe it, and that settles it" kind of family.

I remember being terrified of the vengeful God of the Old Testament. He turned people into pillars of salt. He visited plagues upon the land. He gave Job the worst kind of fits, and I'll never forget the picture of Abraham, his arm upraised as he prepared to plunge a knife into the body of his beloved son.

Oddly though, I never felt that way about the story of Noah. I guess I could have. The idea of a great flood that takes the life of almost every living thing, sparing but a handful to repopulate the planet, could easily terrify any child. But for some reason, it didn't—not then and most certainly not today, now that I'm experienced in dealing with people and with animals. No, I always thought it was a wonderful and elegant story and certainly the second great parable of the Bible. But it seems to me that just about everybody I've ever talked to about it has missed the point of the story entirely—and it doesn't matter if they're Christians or Jews.

Some time ago I had a conversation about this with my dear friend Alvin Katzman. Alvin is Jewish. Not "real" Jewish (that is, conservative) but pretty darn Jewish. ("Darn," by the way, is about the closest a real Methodist ever comes to swearing. Over the entire course of my life, I never once heard my mother use stronger language than "darn"—and when she said "darn," then, boy, you better watch out.) Anyway, he teaches Sunday school at the synagogue. Clearly he must know something about the Bible (or at least, the Old Testament). The conversation started out like this. "Okay Alvin, what's the first great parable of the Bible?" "Easy," he says, "the story of Cain and Abel." "Okay," I say, "and what's the moral of that story?" "Easy," he says, "the moral is that we are all our brother's keeper."

I have to agree with him coming out of the box. Sure, you could say that the moral of the story is that fratricide is bad, but we all know that. The moral is not a negative injunction, like "thou shalt not kill thy brother." It is a positive affirmation of the care we must give to our fellows.

I ask him what the second great parable of the Bible is, and we agree on the story of Noah—but then we hit a snag. When I ask him what the point of the story is, we part ways. Alvin says the point is that God makes a covenant with man that says if we live by the word of God, he will never wreak his vengeance upon us again, and he seals the deal with a rainbow. I say, "Alvin, you have missed the point entirely. Not only do you not get the positive affirmation, but you turn the whole thing back to humans." Okay, I'm not a trained theologian, but it seems so clear to me. "Don't you get it?" I find myself almost shouting: "It's not always about humans!"

Alvin is obviously befuddled. But isn't it pretty obvious that the story is about animals? "Look," I say, "the point is that, second only to our obligation to care for our fellow man, we have an obligation to care for the living things of the earth. It's about stewardship, not sin. We are our brother's keeper, and just as much, we are the keepers of all God's creation."

I have the same conversation with clergy all the time. They say pretty much what Alvin said, I say pretty much what I just said, and interestingly, they usually agree with me. Sort of. They'll say things like, "Yes, it says God gave man dominion over all living things, but the Greek word that we translate as 'dominion' implies responsibility as much as it does ownership. We do, indeed, have an obligation to care for all living creatures." So far, so good— but when you ask them what percentage of their church's budget goes to conservation, the answer will always be the same: nothing. It all goes for human services.

Now, I'm not a particularly nervous person, but if I was in charge of God's spending program, and I was spending *all* of the money on priority one and *nothing* on priority two, I'd be a little concerned that God might take issue with the budget. And, frankly, I'm not sure I'd want to be the one whom God was taking issue with.

I believe that we have a profound, sacred obligation to care for life on Earth. I cannot easily explain it, but I believe it—and that settles it. If I was a theologian, I could provide a richer, more complex (and less succinct) explanation for why this must be true. But I'm an anthropologist. Anthropologists look at the world differently than theologians, and I, for my part, look at the world differently now than I did when I was nine years old.

That difference in perspective was brought home for me not too long ago. Earlier I mentioned that we had an advertising team in to help us roll out a strategy for launching our new WildCare Institute. They seemed like stereotypical ad guys, except they were nicer and they were genuinely interested in what we're trying to do nationally and internationally to save endangered species. They wanted to hear what they called "key words"—words that summarized the whole effort—what we're doing and why. The first word that came out of my mouth was *messianic*. They just stared at me. Indeed, they were every bit as befuddled as Alvin was earlier. So I explained to them that we get the word from the same root as the one for *messiah,* and just like Dan Aykroyd and John Belushi put it in the *Blues Brothers* movie, we do this because "we're on a mission from God."

As a person raised in a very religious home, it is not surprising that I feel a moral obligation to do something that is codified in what I was raised to believe is the most important book ever written. But I wasn't trained to think like a theologian. I was trained to think like an anthropologist, and as a consequence, when I try to explain my convictions in the context of spirituality, I'm like a fish out of water—or a theologically minded chimp who finds himself wondering whether he is his brother's keeper, or his keeper's brother. But perhaps I can explain it by relating it to anthropology, which I do know something about.

As an anthropologist, I often think about the story of Noah in the context of what we call messianic cults. Many messianic cults combine what we might term "conservation" with what we might term "religion," but clearly neither term is perfectly appropriate. As we'll see in a second, messianic cults muddle together elements of both conservation and religion. About the nicest thing you can say about messianic cults is that they're a mess.

Perhaps you're familiar with at least one of the great messianic cults, the

Ghost Dance movement of the Plains Indians. The Ghost Dance started when an Indian prophet named Wodziwob began preaching (perhaps in much the same manner as grandfather did) that all the dead Indians killed by the white men would return on a great train. At the same time, the whites would be swept from the land, leaving all their possessions behind. By the time the movement got to the Sioux, it had expanded to include the promised return of the buffalo, which was by then almost extinct. (This can be viewed in much the same way as conservation is viewed today.) All the Sioux had to do was don a special Ghost Dance shirt and perform the required ceremonies. The kicker was that the shirt would also make them impervious to the US Army's bullets. The Ghost Dance movement ended on December 29, 1890, with the massacre of two hundred Sioux at a place called Wounded Knee, in South Dakota. Tragically, they were wrong about the shirts. At least we eventually got the buffalo back.

We got the buffalo back thanks in large measure to Dr. William Temple Hornaday, who found fame as the first director of the Bronx Zoo. He started

William Hornaday (left) and one of the Bronx Zoo bison ready for transport to Kansas.
WILDLIFE CONSERVATION SOCIETY

his career as the chief taxidermist of the United States National Museum, where, among other things, he shot buffalo for mounting in the museum's collection. He went on to become the superintendent of the Smithsonian's National Zoo and finally moved to New York in 1896 to build the Bronx Zoo.

Hornaday was a passionate conservationist. He argued for, and got passed, the 1911 Fur Seal Treaty, saving the Alaskan fur seal from extinction, and he worked tirelessly on the Federal Migratory Bird Act, which protects migratory birds. But his chief interest was saving the bison. A prolific writer, Hornaday organized and became the first president of the American Bison Society.

The bison certainly needed saving. By 1894 there were only twenty-three animals living in the wild, all in Yellowstone Park. There were a handful of other small groups of plains bison—a small herd in the Palo Duro Canyon in Texas, twelve animals in the Bronx Zoo (which, in 1905, Hornaday would send to an eight-thousand-acre reserve he created in Kansas) plus a grand total of eighty-eight plains bison owned by six private individuals: "Buffalo" Jones, "Sam" Walking Coyote, Charles Alloway, Charles Goodnight, James McKay, and Frederic Dupree. Save for the population in Yellowstone, all of the plains bison alive today are descended from about one hundred animals. The plains bison came that close to extinction. As the Sioux were dancing the Ghost Dance for the plains bison, only twenty-odd animals still roamed free in America. Now, thanks to the conservation efforts of Hornaday and others, there are about two hundred thousand animals in the United States and Canada.

The most fascinating messianic cults might be those found in the islands of Melanesia. One of the earliest examples took place in 1893 at Milne Bay in New Guinea (a place that may well have been the jumping-off point for the brown tree snakes' voyage to Guam). There, a local prophet foretold of great volcanic eruptions and tidal waves that would presage the appearance of huge ships carrying the ancestors as well as "a great abundance of pigs and fruit and all good things." As Marvin Harris dryly notes in his book *Culture, Man, and Nature,* the catch was that all the existing pigs had to be consumed and all European goods had to be abandoned. "After the tidal wave failed to appear," he says, "the colonial officials jailed the prophet to prevent further disturbances." A similar movement started in 1914 on the island of Sabi. This time the ancestors were supposed to come in a steamship laden with money, flour, canned goods, and other valuables. Interestingly, the prayers of the people of Sabi were not just directed to their ancestors. They also prayed to the Christian God, whom they believed had been put in charge of loading the cargo. Then, in Papua, New Guinea, the prophet Evara foretold the arrival of a ship carrying flour, rice, tobacco, and rifles, along with, of course, the dead ancestors. In 1923 a similar cult developed in the New Hebrides.

After World War II, many of the cults began to center on the return of the Americans, who had so mysteriously come to the islands bearing enormous wealth and had just as mysteriously disappeared after the war. Harris calls these new cults "American-oriented revitalizations" and notes that some of these

> placed specific American soldiers in the role of cargo deliverers. On the island of Tana in the New Hebrides, the John Frumm cult cherishes an old G.I. jacket as the relic of one John Frumm, whose identity is not otherwise known. Like many recent cargo movements, the prophets of John Frumm have directed the construction of landing strips, bamboo control towers, and grass-thatched cargo sheds. In some cases beacons are kept ablaze at night and radio operators stand ready with tin can microphones and earphones to guide the cargo planes to a safe landing.

I can't help but find this description poignant. It strikes me, at least in my darker moments, as a twisted metaphor for what I do every day. I mean, maybe we're just fooling ourselves when we build our incredible zoos designed to house these marvelous species. We're thinking that this will work, that we can save them and return them home, that everything will be all right, and that all will be as it once was. But maybe zoos are just like the bamboo airport on the island of Tana—a colossal hoax that we have perpetrated on ourselves. Of course, it's no consolation that most of the messianic cults feature boats prominently—great arks that make everything okay again. People constantly refer to zoos as arks to begin with; I can only hope that the grim hopelessness of the cargo cult isn't our fate.

The vehicles involved in cargo cults often serve double duty; boats, for instance, not only disrupt an indigenous culture, but also are supposed to bring back a prior state of richness. As we've seen, on the Great Plains, it was the railroad that brought the end of the flowering of Native American culture, and so they hoped that the same railroad would bring everything they were losing, or had already lost, back again.

On the islands of the Pacific, though, it was boats. The natives of those beautiful islands were as much at risk as any living thing on an island ecosystem. Like the Guam kingfisher, or the Partula snails, or the exquisitely rare Jamaican ground iguana, when they are threatened with disaster, they simply have nowhere to go. The lives of the islanders, although not ended, were transformed forever by the arrival of "civilization." They had to believe that the same boats that brought the destruction of their former way of life could also bring the return of their way of life. It wasn't until after World War II that planes became the simultaneous metaphor for destruction and the hope for resurrection. The people of those islands wanted the world around them restored to the way it was before the engines of the modern world began to use their islands to fuel the industrial machine.

Speaking as an anthropologist, I believe that the people of the Old Testament wanted much the same as the people of Oceania. The world of the Chosen People was transformed by the Neolithic Revolution, a revolution that started in the Middle East and that they, or their ancestors, were most certainly a part of. Perhaps they, too, saw how the juggernaut of civilization was altering forever their natural world. Perhaps, like the people of the Pacific, they dreamed of a great boat that would bring their natural environment back to them after the waters had cleansed the land. Maybe the Chosen People were the first cargo cult in recorded history, documenting the dream of the return of their old world while struggling to survive in their new one.

The stories of cargo cults don't always have to be strange and bizarre fantasies of native Pacific Islanders, relayed to us indirectly via modern anthropologists. The story of Noah might indeed be a wonderful story handed down to us by the people themselves, and if so, it is a richer, deeper story for its directness. It tells us that, no matter how bad things are in our world, there is hope that the planet can be renewed and that the living things can be saved. By using human ingenuity—the same kinds of early engineering skills that had changed the world of the Chosen People—to build his three-hundred-cubit boat, Noah transformed the engines of change into a vehicle of salvation. The story of Noah is the story of hope, not destruction. It is less about what will happen if we continue to do wrong, and more about what could and should happen if we take on the individual and collective responsibility to do right.

So maybe people who work in zoos are less like the people of Tana, who build landing strips for planes that will never come, and more like those two bicycle mechanics from Dayton, Ohio, who designed a machine that, on

December 17, 1903, flew some 852 feet through the air. The Wright brothers, along with a handful of other visionaries, changed the world forever. Even a tiny group of people—Noah's family, for instance, according to the Bible—can make an extraordinary difference. Margaret Mead used to say it all the time. "Never doubt," she would say, "the ability of a small number of dedicated individuals to change the world. Indeed, it is the only thing that ever has."

Still, I doubt. I used to say that the day I gave up hope for the environment—wrote its epitaph, decided it was useless to try to save the many species that clung to life by slim threads—was the day that I should write my own epitaph. The problem with that is, it doesn't leave a lot of wiggle room. Giving up on the lives of others would mean, by definition, giving up on my own. I am not yet ready to do either.

But we are losing. And just as many species are losing their ability to bounce back from the afflictions that humanity has lent them, so I am losing my resilience. I do not doubt the ability of a small group of dedicated individuals to change our world. I doubt sometimes that I can be one of them.

When I first started to work in zoos, I found two things that tempered the joy of working in these wonderful places. First, the people who work in zoos feel that everything in their care will live forever. Of course, they know in their heads that the exact opposite is true. But in their hearts they feel that what they care for will never die. Every time something they care for does, they feel an anguish that they never become used to. Second, they feel that there are never enough resources to win the war on extinction—never enough time, never enough money. And they feel a compelling urgency. We all do. As Gro Harlem Brundtland (a former prime minister of Norway) put it at one of the Trondheim conferences on biodiversity, "The library of life is burning and we do not even know the titles of the books."

These two burdens will never go away, and we will never become inured to them. We have no choice but to continue to care and continue to try. We will not, perhaps, halt the loss of biodiversity on our Earth in its entirety, but we can make a difference. We're obligated to do so. That is what I must tell myself. That is what I must convince you of.

I am a scientist. But science, says Leo Tolstoy, cannot answer the central questions of life: "What shall I do?" "How shall I live?" Science can tell us many things. It gives us extraordinary insights into how our world functions. It opens our eyes to worlds smaller than an atom—places that we cannot

even see—and to worlds so large that we cannot possibly comprehend them—
a universe unimaginably vast. Science can tell us much about the web of life
that surrounds us and sustains us. It can tell us how our actions, for better or
for worse, might change our planet, and it can give us insights into how
those changes will affect us as individuals, as communities, and as a species.
But it cannot tell us what we should do; it cannot tell us how we should live.
In attempting to answer those questions, we must look to religion, ethics,
philosophy; we must ultimately look within. As individuals, we must each
decide what we shall do and how we will shall live. And if enough of us decide to act differently, then our world will change.

People often think of zoos as the modern version of Noah's ark. But zoos
are most certainly not arks; as Eric Miller always says, "we're more like lifeboats." We can save only a *tiny handful* of species in zoos; Noah had two of
everything. Sometimes, we may look a little like the ark in that we help to
keep afloat populations that are extraordinarily small—often only a few hundred and sometimes less than a hundred. It's better, I suppose, than being
down to the biblical "two by two," but not much.

Even if we are merely lifeboats, it really isn't up to zoos alone to preserve
the wealth of plants and animals on our Earth. Instead, it is up to *all of us*. I
believe that the cumulative effect of individual actions can result in great
changes—that you and I, working together, can arrest and even turn back the
tide of extinction. Therein lies the special mission of our zoos. I wish I could
remember who said this to me. Whoever it was, I believe him: "It's hard to
walk around a good zoo without caring, deeply, about whether this miraculous wealth of lovely, peculiar, creepy, unfathomable creatures survives or
perishes."

No, zoos are most definitely not arks, but our planet is. Our planet con-
tains all of the life that we have, and it floats in the vastness of space just as
Noah's craft floated on the infinite waters. Our planet is an ark, and the living
things of the planet are our treasure. Our obligation is every bit as sacred and
profound as Noah's. We all must do everything we can to preserve life on
Earth. It is our sacred obligation to our selves and to one another.

Like it or not, the world is an ark. Like it or not, we are all sailing with
Noah.

Epilogue

Sitting on this beautiful beach, looking out over the Caribbean, it is hard to imagine that there is anything amiss with our world. The water is five completely different hues of blue, stretching in wide bands from the deep azure of the horizon, to the striking teal waters just off-shore, then finally blending into the pure white of the waves washing up on the beach. In a while, I'll put on my mask and fins and dive the shallow reef not far from here. I won't see anything I haven't seen on any other of our annual pilgrimages to this place, but it will be magical all the same.

This is supposed to be a vacation, and I feel a little guilty for working when I should be relaxing. But my wife, ever understanding, just shrugs. "Some people read books on vacation," she says, "some people write them. Whatever floats your boat." She, of course, knows the working title of this book and I figure she's teasing me a little, but I keep writing all the same.

We're here because ten years ago my wonderful friend, Alvin Katzman (who would probably refer to himself as "the Jew in Chapter 18"), introduced us to this little bit of paradise—another small favor that I will never be able to repay. It's a good place to finish a book, though, especially since it's just twenty degrees in St. Louis. There's a moral in there somewhere. Perhaps it's just that everything is relative. The world is in bad shape, but it could be much worse. Many dedicated people are devoting their lives to making it better. In many places we are winning. Sometimes the victories are small ones, but they are victories nonetheless.

I recently returned from a trip to London to work on the ISIS project with our European partners. While there, I met with Paul Pierce-Kelly, one of the original founders of the *Partula* snail project, at the Zoological Society of London. He was, as always, bubbling with enthusiasm, showing me the computer program he developed to track *Partula* genetics. He designed it partly to help with a problem that ISIS has—tracking the genetics of a population when we don't actually know which animal bred with which. It is a tricky problem, but Paul's solution was ingenious, and it will no doubt be incorporated into the final version of the ISIS software. Paul brought me up to speed on the world's smallest reserve, the one for the partula snails on Moorea. He has worked with a field scientist to fix the technical problems of keeping the carnivorous snails out and is now planning a whole series of microreserves. Before I leave, he gives me an old faded brown envelope covered with Ulie Seal's distinctive scrawls. It is the envelope on which Ulie sketched out the *Partula* snail Species Survival Plan. Paul asks me to take the envelope back to America for Ron Goellner to sign. Ron's is the last signature Paul needs. After Ron signs, everyone who was present when Ulie established the first SSP for an invertebrate will have autographed a very historic, albeit plain, piece of scrap paper.

Many things have not changed since I began writing this book. The very sick elephants of my early days at the Indianapolis Zoo continue to thrive, and their appetites have never flagged since the day Debbie and my wife began tossing them soggy bread balls. Pete Hoskins continues to run the Philadelphia Zoo, and the gorillas of Chicago's Brookfield Zoo continue to inhabit Tropic World. The young boy who fell into the gorilla exhibit still occasionally visits, but not with his mother and older brother and sister. The episode was terribly traumatic for them; his grandparents bring him instead. Rob Shumaker and his orangutans have left the National Zoo, though. There is a new facility in Ames, Iowa, that has been designed to house primates that are involved in experiments on learning. Oddly, the consultant who worked for years on this project and, when it neared completion, was charged with hiring the primate center's first director, was James Abruzzo, the same man who brought me to St. Louis.

There have been defeats and setbacks, though. Amali, the first baby African elephant conceived via artificial insemination, died on June 3, 2003. She had a severely impacted bowel and underwent emergency surgery. It looked like it would be the first successful surgery ever performed on an African elephant but, after appearing to be on the road to a full recovery, she succumbed to a secondary infection. Judy Gagen once told me that she felt that we learned as much from her death as we did from her birth, but it is still painful.

Lena, the Siberian tiger who hated my guts, died of cancer. In another one of those weird twists of fate, one of her offspring now lives in the Saint Louis Zoo. Unlike her mother, she adores me, and sometimes I slip behind the scenes to run with her along the edge of her huge enclosure. It is humbling for me, though. I can run as fast as I can, and she barely has to break into a trot in order to race past me.

My precious kingfishers have also suffered a setback. At the beginning of 2004, things were actually looking pretty good. The overall population size had shot up to about seventy birds, with four chicks born at our zoo last year. More importantly, last summer, our vet, Dr. Randy Junge, traveled to Guam with several of the delicate little creatures. He felt that it would be possible to breed them more successfully in a protected aviary in Guam and, if successful, we were looking to release them in a sixty-acre snake-proof habitat that the Air Force was planning to build on Andersen Air Force Base.

Andersen AFB has 24,803 acres of tropical forest—the only remaining wild habitat on the island of Guam. If we are to save the endangered birds of Guam, Andersen must play a pivotal role. But last year the Bush administration proposed, and Congress approved, an order that exempts military facilities from the "critical habitat" provisions of the Endangered Species Act. Like you, I never even noticed.

But I did notice when, late last year, the Air Force announced plans to expand the base, reducing the 26,803 acres of critical habitat down to 376 acres. In other words, government biologists have been told to expect that most of the forest will be cut down to (quite literally) pave the way for a $1 billion to $2 billion expansion of the base.

When the government advocated exempting the military from the critical habitat provision of the law, it argued that we could trust our government in general and our military in particular to protect the environment. In December 2004, the military cancelled their plans to build the snake barrier. Now they are planning to cancel any hope that our beautiful kingfishers might have of ever living wild and free. Forgive me, but if I can no longer trust them to protect the environment of the Micronesian kingfisher, then I wonder if I can trust them to protect the place where I live. These two go together. They are inextricably interwoven. Oren Lyons, a chief of the Onondaga Nation, said it best: "You cannot have peace in this world if you continue to make war on Mother Earth."

Still, there is hope, and we take our victories where we find them. Our zoo continues to flourish and was recently voted the top zoo in America. The work of our new WildCare Institute grows in scope and importance all the time. Not all is right with the world, but not everything is wrong, either.

So, that brings me to today and this beach. None of these stories have really ended, though. They will go on. And there are so many more stories to tell. Perhaps I will have another opportunity to tell you some of them. I would like that, if only because I learned so much in researching and writing this book. I suppose that, the more you learn, the more you change. I know that, as a result of looking into all of these different stories more fully, my beliefs about conservation have shifted in many ways, both great and small. For example, I know I said in the introduction to this book that we must speak for all living things. I still believe that this is true, but I believe, now more than ever, that it is more important that we *act* for all living things.

In the movie *The Shawshank Redemption,* the lead character, in a note to his best friend, talks of hope. He says, "Hope is a good thing, perhaps the best of things." When I began this book, I would have agreed with him; I no longer do. Hope is passive. It lacks the obligation and the incentive to act. Hope is what we have left when we cannot control our own destiny.

I now believe that we should talk less about hope and more about another quality—the quality of mercy. If wild things and wild places could speak to us, they would *not* ask us to share with them our hopes. Instead, they would ask us to show them our mercy. Only by showing mercy can we give them hope.

Terry Tempest Williams, in a piece entitled "Wild Mercy," says, "The eyes of the future are looking back at us, and they are praying for us to see beyond our own time. They are kneeling with clasped hands that we might act with restraint, leaving room for the life that is destined to come. To protect what is wild is to protect what is gentle. Perhaps the wilderness we fear is the pause within our heartbeats, the silent space that says: 'We live only by grace.' Wilderness lives by this same grace. We have it in our power to create merciful acts."

The word *mercy* might strike many people as odd, or even offensive. Wild things have done us no harm. "Why," you might ask, "is he talking about granting mercy to living things? Who are we to sit in judgment over the creatures of the earth?"

I think now of the ancient gods—the gods of the Greeks and the Romans, the gods of the Egyptians and the Hindus, the gods of the ancient Americans.

Those gods remind me of ourselves. They were noble at times and sometimes base. They were kind and sometimes cruel. They interceded in the natural world, sometimes on a whim, sometimes out of spite, and sometimes to right a terrible wrong. They could be capricious or thoughtful, benign or neglectful.

As a species, we have become much like them. We can influence the very character of our planet, of our oceans and seas, of our lands, and of the very air that we breathe. We have powers even beyond them. The fate of so many species rests squarely in our hands. Like the ancient gods, we can destroy or sustain. People both ancient and modern have beseeched their gods and called upon their most merciful natures. Now we must accept as a species that awesome responsibility as our own.

Wild things *are* speaking to us. They are saying, "Grant us mercy." We are the only species on this planet that can, for better or worse, intentionally change the very character of the natural world. We can do so thoughtfully, with grace and compassion, or we can do so thoughtlessly, with avarice and ruthlessness. All of us have it in our power to show mercy. Let this be the quality that drives our hearts and guides our minds. Let this be the quality that informs our decisions and forms the very core of our values. Let this be the single criterion that we hold most dear. Let us always ask ourselves, "Is this how we can best show mercy toward other living things?" "Is this how we can best show clemency toward life on Earth?" Let us do more than hope. Let us act mercifully and as one.

Paradise Island,
The Bahamas
January 21, 2005

Selected References

Introduction
Beston, Henry. *The Outermost House.* 1928. Reprint, New York: Henry Holt, 1992.

1. Escargot, Anyone?
Brues, Alice. "The Spearman and the Archer: An Essay on Selection in Body Build." *American Anthropologist* 61, no. 3 (June 1959): 457–69.

Leopold, Aldo. *A Sand County Almanac.* San Francisco: Sierra Club Books, 1974.

Lyman, R. Lee. *White Goats, White Lies: The Abuse of Science in Olympic National Park.* Salt Lake City: University of Utah Press, 1998.

Todd, Kim. *Tinkering with Eden: A Natural History of Exotics in America.* New York: Norton, 2001.

3. The Tragedy in Philadelphia and the Miracle in Chicago
DeLeon, Clark. *America's First Zoostory: 125 Years at the Philadelphia Zoo.* Virginia Beach, Va.: Donning, 1999.

Friend, Tim. *Animal Talk: Breaking the Codes of Animal Language.* New York: Free Press, 2004.

4. When a Butterfly Sneezes
Wilson, Edward O. *Biophilia.* Cambridge: Harvard University Press, 1984.

5. The Telemetric Egg
Audubon, John James. *The Audubon Reader: The Best Writing of John James Audubon.* Ed. Scott Russell Sanders. Bloomington: Indiana University Press, 1986.

Cokinos, Christopher. *Hope Is the Thing with Feathers.* New York: J. P. Tarcher/Putnam, 2000.

Fuller, Errol. *Extinct Birds.* New York: Facts on File, 1987.

6. A Virus among Us
Ralls, K., K. Brugger, and J. D. Ballou. "Inbreeding and Juvenile Mortality in Small Populations of Ungulates." *Science* 206 (1979): 1101–3.

"West Nile Fever: A Medical Detective Story." American Museum of Natural History *BIOBulletin,* www.amnh.org/education/resources/bulletins, accessed June 28, 2005.

7. Lions and Tigers and Bears
Bircher, Steve. "Barely Saved from a 'Grizzly' End." *Zudus* 5, no. 5 (September/ October 1991): 1.

8. Golden Tamarins and Ferrets with Black Feet

Asa, Cheryl. "Turning the Tide: Channel Island Fox Recovery." *Zudus* 17, no. 2 (March/April 2002): 4.

"Captive Propagation and Reintroduction: A Strategy for Preserving Endangered Species?" *Endangered Species Update* 8, no. 1 (special issue) (November 1990).

9. Ralph Neds I Have Known

Linden, Eugene. *The Octopus and the Orangutan: New Tales of Animal Intrigue, Intelligence, and Ingenuity.* New York: Dutton, 2002.

11. The Stradivarius of Birds

Corrigan, Patricia. *Wild Things: Untold Tales from the First Century of the Saint Louis Zoo.* St. Louis: Virginia Publishing, 2002.

13. Skukuza

Luard, Nicholas. *The Wildlife Parks of Africa.* London: Michael Joseph, 1985.

14. The Panda Wars

Kleiman, Devra. "Giant Pandas." *AAZPA Communiqué,* February 1993, 8–12.

Maple, Terry L. *Saving the Giant Panda.* Atlanta: Longstreet Press, 2000.

Rubel, Alex, Matthew Hatchwell, and James MacKinnon. *Masoala: The Eye of the Forest: A New Strategy for Rainforest Conservation in Madagascar.* Zurich: Zoo Zurich, 2003.

Schaller, George B. *The Last Panda.* Chicago: University of Chicago Press, 1993.

15. This Would Be a Nice Place—If It Weren't for the Visitors

American Zoo and Aquarium Association. *Standards for Elephant Management and Care.* Adopted March 21, 2001; updated May 5, 2003.

Croke, Vicki. *The Modern Ark: The Story of Zoos—Past, Present, and Future.* New York: Scribner's, 1997. (Terry Maple quote from page 171.)

Hyson, Jeffrey. "No Business Like Zoo Business: Research Is Fine but Show Us the Animals." *Washington Post,* May 11, 2003, B1.

16. The Problem with PETA

American Zoo and Aquarium Association. Animal Welfare Committee. Welfare Metrics Subcommittee. (Definition of *animal welfare.*)

Barnes, Fred. "Animal-Rights Activists Now Threaten Medical Research." *Vogue,* September 1989, 542.

Clark, Ward M. *Misplaced Compassion: The Animal Rights Movement Exposed.* Lincoln, Neb.: Writers Club Press, 2001.

McCabe, Katie. "Who Will Live, Who Will Die?" *Washingtonian,* August 1986, 113-57.

Regan, Tom. "Are Zoos Morally Defensible?" In *Ethics on the Ark: Zoos, Animal Welfare, and Wildlife Conservation,* ed. Bryan G. Norton, Michael Hutchins, Elizabeth F. Stevens, and Terry L. Maple. Washington, D.C.: Smithsonian Institution Press, 1995.

Singer, Peter. *Animal Liberation: A New Ethics for Our Treatment of Animals.* New York: New York Review, distributed by Random House, 1975.

Specter, Michael. "The Extremist: The Woman behind the Most Successful Radical Group in America." *New Yorker,* April 14, 2003, 57.

Terhorst, Cheryl. "Where Would We Be without Animals?" *Chicago Daily Herald,* March 1, 1990, sec. 3.

Wright, Robert. "Are Animals People Too? Close Enough for Moral Discomfort." *New Republic,* March 12, 1990, 20.

17. Zoos of the Future

Conway, William. "The Changing Role of Zoos in the 21st Century." In *Proceedings of the 54th World Zoo Organisation Annual Conference.* Pretoria, South Africa, October 17-23, 1999.

Gardner, Howard. *Frames of Mind: The Theory of Multiple Intelligences.* New York: Basic Books, 1983.

Hoerr, Thomas R. *Becoming a Multiple Intelligences School.* Alexandria, Va.: Association for Supervision and Curriculum Development, 2000.

18. Sailing with Noah

Geist, Valerius. *Buffalo Nation: History and Legend of the North American Bison.* Stillwater, Minn.: Voyageur Press, 1996.

Harris, Marvin. *Culture, Man, and Nature: An Introduction to General Anthropology.* New York: Thomas Y. Crowell, 1971.

Epilogue

Williams, Terry Tempest. "Wild Mercy." In *Red: Passion and Patience in the Desert.* New York: Pantheon Books, 2001.

Index

308   Index

208–10; Javan, 205; in Kruger National
    Park, 206–7; PETA on importation of,
    260–62; white, 205, 260–61
Riemvasmaker people, 201
Rimba (sun bear), 250
Rita and Jinga Gula (orangutans), 36–37
Roger Williams Park Zoo (Rhode Island),
    135, 273, 281
Roosevelt, Franklin, 17
Royal flycatcher, 99
Rubenstein, Dan, 71–72
Ruhe, Heinz, 214–15

Sabie Reserve. See Kruger National Park
Sahelo-Saharan region, 269, 280
Saint Louis Zoo: animal escapes at, 145–52,
    155–59; animal shows at, 183–85, 262;
    bears at, 121, 123, 248–49; carousel at,
    179–80, 185; cheetahs at, 145–46,
    149–52; chimps at, 139–44, 157–59,
    262; community education in New
    Guinea by, 111; in conservation efforts on
    Madagascar, 226–27, 270, 275; conserva-
    tion work by, 90, 222, 263, 271–72; ele-
    phant facilities at, 243; finances of,
    177–80; grizzly bears at, 121, 123; king-
    fishers at, 177, 296; lions at, 146–47;
    mixed mission of, 233–34; obligation to
    animals once under its care, 265–66; pan-
    das at, 214–19; PETA and, 250, 259,
    262–63, 266; research by, 8, 63, 70–73,
    227; sponsoring trips, 189–97; sun bears
    at, 248–49; tamarins in, 132; tigers at,
    296; visitors' reasons for coming to,
    234–35; waterfowl at, 68; WildCare
    Institute at, 271–81, 273–81
Saint Louis Zoo Friends, 86, 272
Salmon, Daniel, 23
Salmonella, in elephants, 23–25, 27–30,
    32–33
Sambia people, 31
Samson (gorilla), 42
San Diego Zoo, 224, 280
Sandridge, Suzanne, 20
San Francisco Zoo, 226
Santa Barbara Zoo, 137
Schaller, George, 212, 220, 221, 223
Schieffelin, Eugene, 11
Schmitt, Dennis, 172–73
Schmitt, Ed, 221
Schnell, Keith, 154
Schwendermann, Matt, 38
Scimitar-horned oryx, 89

Seal, Ulysses S., 19, 61–62, 131, 295
Sea lion shows, at Saint Louis Zoo, 185
Sekaye (lion), 118
Sepilok Rehabilitation Center, 247
Seyfried, Alice, 280–81
Shamfa (lion), 116–17
Shanto (cheetah), 152
Shrikes, San Clemente, 139
Shumaker, Rob, 52, 295
Siberia, tiger habitat in, 118–19
Sikhote-Alin International Biosphere Reserve
    (Siberia), 118–19
Simmons, Lee, 53, 160
Singer, Peter, 253
Skeleton Coast, Namibia, 202–4
Skukuza, 204
Smith, Floyd Tangier, 214
Smith, Lorraine, 140–41
Smoke (chimpanzee), 139
Snails: Achatina, 18–19, 21; E. rosea, 19, 21;
    partula, 17–21, 77, 105, 295
Snakes: brown tree, 176–77; mountain
    vipers, 70–71, 113; Mozambique spitting
    cobra, 204–5; spitting cobra escape,
    147–49
Society Islands, partula snails in, 18
Sophie (elephant), 24, 26–27
Sossusvlei, Namibia, 202
South Africa. See also Kruger National
    Park
South America, mass extinctions in, 77
Sparrows, house, 10–11, 13–15
"Spearman and the Archer" (Brues), 15
Species Survival Plans (SSPs), 82, 85; for
    African lions, 116, 118; avoiding inbreed-
    ing in, 89–90; for black-footed ferrets,
    127–28; for chimps, 265; lack of, for pan-
    das, 221; for sun bears, 247–48; in wider
    survival planning, 90–91
Starlings, 10–11
Stehr, John, 260
Stevens, Jane, 273
Stevenson-Hamilton, James, 207–8
Stewart, Ken, 264–65
St. Louis Museum of Science and Natural
    History, 3–4
St. Louis Science Center, 4, 81
St. Louis University, 276
St. Louis Zoo. See Saint Louis Zoo
Stuemke, Chet, 155–56
SUNY–Stony Brook, 226, 270
Sussman, Robert, 73
Swakopmund, Namibia, 201–2

About the Author

Jeffrey P. Bonner is the President and CEO of the Saint Louis Zoo and Past President of the Indianapolis Zoo and White River Gardens. He is a Burgess Fellow, Travelling Fellow, Fulbright Scholar, President's Fellow, and a recipient of the National Research Service Award.